Reliability and Radiation Effects
in Compound Semiconductors

Reliability and Radiation Effects in Compound Semiconductors

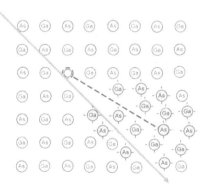

Allan Johnston

Jet Propulsion Laboratory
California Institute of Technology, USA

NEW JERSEY · LONDON · SINGAPORE · BEIJING · SHANGHAI · HONG KONG · TAIPEI · CHENNAI

Published by

World Scientific Publishing Co. Pte. Ltd.
5 Toh Tuck Link, Singapore 596224
USA office: 27 Warren Street, Suite 401-402, Hackensack, NJ 07601
UK office: 57 Shelton Street, Covent Garden, London WC2H 9HE

British Library Cataloguing-in-Publication Data
A catalogue record for this book is available from the British Library.

Book cover image: Artist's rendition of the Cassini spacecraft near Saturn (courtesy of the National Aeronautics and Space Administration, NASA)

RELIABILITY AND RADIATION EFFECTS IN COMPOUND SEMICONDUCTORS

Copyright © 2010 by World Scientific Publishing Co. Pte. Ltd.

All rights reserved. This book, or parts thereof, may not be reproduced in any form or by any means, electronic or mechanical, including photocopying, recording or any information storage and retrieval system now known or to be invented, without written permission from the Publisher.

For photocopying of material in this volume, please pay a copying fee through the Copyright Clearance Center, Inc., 222 Rosewood Drive, Danvers, MA 01923, USA. In this case permission to photocopy is not required from the publisher.

ISBN-13 978-981-4277-10-5
ISBN-10 981-4277-10-X

Printed in Singapore.

Preface

The motivation for writing this book stems from many years of involvement in the radiation effects field, as well as concerns about the applicability of conventional reliability models to long-duration space missions where component failure can potentially bring the mission to a premature end. It was also influenced by the difficulty of dealing with the wide range of materials used in compound semiconductors, as well as in understanding the very different types of electronic devices that have been developed for those materials.

Reliability and radiation effects are both important for spacecraft. Both areas are far more complex for the semiconductor technologies available today compared to those found on earlier spacecraft. This is particularly true for devices that use compound semiconductors, which have not been studied as thoroughly as those using silicon.

Reliability and radiation effects both require a knowledge of semiconductor physics and device structures that goes beyond the minimum level required for circuit design. The purpose of this book is to discuss both topics in a cohesive manner, pointing out specific areas where there are interactions between the two disciplines. The book begins with background material on semiconductor physics and devices, emphasizing topics such as heterostructures and modulation-doped field-effect transistors that are less familiar to many engineers and designers. This is followed by three chapters on reliability, starting with basic reliability concepts and models, which are then applied to specific types of compound semiconductor transistors and optoelectronics.

The last half of the book discusses radiation effects. Chapter 8 reviews radiation environments near the earth, along with a brief discussion of terrestrial radiation environments. The next chapter discusses the physical processes involved when radiation interacts with solids and semiconductors, providing the background for later chapters that discuss radiation damage in electronic and optoelectronic devices. There is a separate chapter on optocouplers because they use both silicon and compound semiconductor devices, with many different design options that influence radiation sensitivity.

The last chapter discusses the effects of single energetic particles on compound semiconductors. Although this usually produces short-duration transient effects, in some cases devices are permanently damaged by such particle strikes.

Compound semiconductors evolve rapidly, using new materials and device concepts to take advantage of the unique properties of the many different materials that are available to improve device performance. Although this is beneficial from the standpoint of performance, it increases the difficulty of writing a book of this type. Hopefully the emphasis on the underlying physics and mechanisms will make it useful for many years, providing the background and level of understanding that is needed to apply the disciplines of reliability and radiation effects to more advanced devices.

Acknowledgments

Many colleagues and organizations have helped me along the way. I owe a particular debt of gratitude to the Department of Physics at the University of Washington, Seattle, Washington, for motivating my study of physics and providing a solid technical background. I have also benefited from participation in conferences on radiation effects that are sponsored by the IEEE Nuclear and Plasma Sciences Society, and interactions with numerous contributors and conference organizers.

I would like to acknowledge Dr. Robert Caldwell and Dr. Itsu Arimura of the Boeing Aerospace Company, who provided inspiration and guidance early in my career. Mr. Sam Kayali, Dr. Charles Barnes and Dr. Henry Garrett are among the many colleagues at the Jet Propulsion Laboratory that have contributed to my success.

Finally, I would like to thank my parents, Harold and Margaret Johnston, my uncle, John W. King, and my wife Miriam.

Allan Johnston
Pasadena, California

Contents

Preface v
Acknowledgments vii

1   Introduction 1
   1.1   Basic Considerations 1
   1.2   Some Specific Examples 2

2   Semiconductor Fundamentals 9
   2.1   Fundamental Concepts 9
   2.2   p-n Junctions 23
   2.3   Fabrication and Material Issues 33
   2.4   Power Applications 36
   2.5   Summary 38

3   Transistor Technologies 41
   3.1   Junction Field-Effect Transistors 42
   3.2   Modulation-Doped Field-Effect Transistors 47
   3.3   MOS Transistors 55
   3.4   Bipolar Transistors 58
   3.5   Noise 68
   3.6   Summary 71

4   Optoelectronics 75
   4.1   Critical Semiconductor Properties for Light Emission 75
   4.2   Material Considerations 81

	4.3	Light-Emitting Diodes	85
	4.4	Laser Diodes	91
	4.5	Detectors	106
	4.6	Summary	112
5	Reliability Fundamentals		117
	5.1	Reliability Requirements	118
	5.2	Acceleration Mechanisms	126
	5.3	Basic Failure Mechanisms	132
	5.4	Analysis of Reliability Test Data	135
	5.5	Summary	140
6	Compound Semiconductor Reliability		143
	6.1	MESFETs and HFETs: Mature Technologies	144
	6.2	GaAs Heterojunction Bipolar Transistors	155
	6.3	SiGe Heterojunction Bipolar Transistors	158
	6.4	Wide Bandgap Semiconductors: SiC and GaN	160
	6.5	Summary	166
7	Optoelectronic Device Reliability		171
	7.1	Basic Considerations	171
	7.2	Reliability of Light-Emitting Diodes	176
	7.3	Laser Diode Reliability	183
	7.4	VCSELs	194
	7.5	Tunable and Frequency Stabilized Lasers	196
	7.6	Optical Detectors	199
	7.7	Summary	201
8	Radiation Environments		205
	8.1	Particle Types	205
	8.2	Radiation Environments Near the Earth	213
	8.3	Energy Distributions in the Earth's Trapped Belts	215
	8.4	Radiation Environment in a Geosynchronous Orbit	217
	8.5	Protons from Solar Flares	218
	8.6	Galactic Cosmic Rays	223
	8.7	Heavy Particles in Solar Flares	225

	8.8 Terrestrial Environments	229
	8.9 Summary	232

9 Interactions of Radiation with Semiconductors 235
   9.1 Fundamental Interactions 235
   9.2 Effects of Damage on Semiconductor Properties 246
   9.3 Radiation Effects in Heterostructures 253
   9.4 Energy Dependence of Displacement Damage 254
   9.5 Summary 259

10 Displacement Damage in Compound Semiconductors 263
   10.1 JFETs 265
   10.2 HFETs 266
   10.3 Advanced Bipolar Transistors 269
   10.4 Wide-Bandgap Devices 273
   10.5 Summary 279

11 Displacement Damage in Optoelectronic Devices 283
   11.1 Light-Emitting Diodes with Amphoteric Doping 284
   11.2 Heterojunction LEDs 292
   11.3 Edge-Emitting Laser Diodes 296
   11.4 VCSELs 303
   11.5 Photodetectors 305
   11.6 Summary 307

12 Radiation Damage in Optocouplers 311
   12.1 Introduction 311
   12.2 Damage in Basic Phototransistor Optocouplers 313
   12.3 Optocouplers with High-Speed Internal Amplifiers 324
   12.4 Optocouplers with MOSFET Output Stages 327
   12.5 Summary 330

13 Effects from Single Particles 333
   13.1 Basic Concepts 334
   13.2 Single-Event Upset in Logic Devices 339
   13.3 Optocouplers 343

13.4 Optical Receivers 348
13.5 Single-Event Upset Effects from Neutrons 351
13.6 Permanent Damage from Particle-Induced Transients 352
13.7 Summary 355

Index 357

Chapter 1

Introduction

1.1 Basic Considerations

Silicon technology continues to dominate the semiconductor industry. Numerous breakthroughs have resulted in the ability to produce extremely large scale devices on a single die. Many reliability problems have been solved along the way, but new reliability mechanisms are present in compound semiconductors that are not fully addressed by the methods used to increase reliability in silicon.

The most critical difference between silicon and compound semiconductors is in the properties of the basic materials. Silicon has many advantages, including low defect density, high material strength, and good thermal conduction. On the other hand, compound semiconductors involve a host of different materials with widely different properties. Some can only be produced in thin layers, and must be grown on substrates with different lattice spacing. With the exception of SiGe, none have the low defect density of silicon. Moreover, new reliability problems such as gate sinking and current collapse have emerged for which conventional approaches to reliability are not necessarily effective. A far better understanding of material properties is needed in order to understand reliability and radiation effects in these types of semiconductors.

Many of the device structures involve advanced concepts that are quite different from the comfortable world of bipolar and metal-oxide-semiconductor (MOS) transistors made with silicon. For compound semiconductors, one is immediately forced to deal with heterostructures, quantum confinement, and "2DEG" gas concepts, as well as materials with different bandgaps and physical properties. It can be difficult to sort through these ideas and determine how the more familiar ideas of reliability apply to such structures.

Considerable progress has been made in the development of compound semiconductors during the last 15 years. Early work concentrated on GaAs MESFET technology, primarily for radio frequency (RF) applications, but more recent emphasis has been on developing other devices and materials that provide even better high-frequency performance. Heterojunction bipolar transistors (HBTs) have been developed that use InP-based or SiGe, with extremely high bandwidth and reduced switching time. This has made it possible to design individual transistors with bandwidths above 500 GHz. There has also been extensive progress in developing RF switching modules, where high-frequency performance and high output power are the main requirements. Applications of compound semiconductors can be divided into three basic areas:

(1) Low and moderate power logic and analog applications that take advantage of the very high operating frequency provided by compound semiconductors, particularly SiGe and InP.
(2) RF power modules that require very high power density. Although many such applications can be met using GaAs MESFETs, new materials such as GaN can operate at higher temperature and voltage.
(3) Optoelectronic devices, particularly LEDs and laser diodes. The development of new materials and fabrication methods has expanded the range of wavelengths for optoelectronics, and this continues to be an active development area.

There are other specialized applications that do not fit within these categories, including those involving high voltage or high temperature.

1.2 Some Specific Examples

Two examples are discussed below that help to establish the context for the material in the book.

1.2.1 Reliability of advanced GaN transistors

A recent paper discussed progress in the development of GaN transistors intended for high-power, high-frequency applications as well as high voltage [1]. Figure 1-1 compares the specific on resistance of various GaN devices with their silicon counterparts as a function of breakdown voltage; the dashed lines show theoretical limits for the two materials. The silicon devices are somewhat better than the limiting value, but do not perform as well as the GaN devices. Even at lower voltages, GaN has a clear advantage. At higher voltages, GaN devices provide an on resistance that is more than an order of magnitude lower than those using silicon.

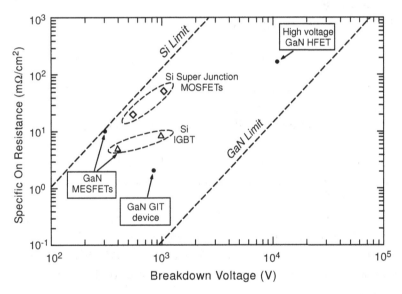

Fig. 1-1. Comparison of various GaN transistors for high-voltage and high-power applications [1]. © IEEE 2008. Reprinted with permission.

Taken at face value, GaN is the obvious choice for applications above 300 V. In addition to lower values of on resistance, the transconductance is higher than for the silicon devices. However, these are complicated devices. The GaN GIT (gate injection transistor) is fabricated on a silicon substrate, which has a different lattice constant. It also has a highly nonlinear transconductance curve, which could be a problem in some circuit applications.

The high-voltage device, capable of operating at 10,000 V, is fabricated on a sapphire substrate, which also has a lattice mismatch with GaN. This not only affects defect density, but also limits power dissipation because of the low thermal conductivity of sapphire. A new material, AlN, is used as a passivation layer to increase the breakdown strength.

Overall, these are impressive characteristics, but the two devices are not yet in production. A relatively sophisticated knowledge of material and device properties is required in order to determine whether either structure is near the point where it could be produced, or represents just another advance at the laboratory level.

There are several potential reliability issues:

(1) Both devices use other substrate materials, with different lattice matching.
(2) Thermal conductivity is much lower for the sapphire substrate, limiting power dissipation.
(3) The gate structure of the GIT device is new, and it is uncertain how it may affect reliability.
(4) The properties of AlN are not well known. It was the key breakthrough in achieving the high operating voltage, but could affect reliability.
(5) The operation of the GaN HEMT depends on extremely thin layers, as well as on the presence of a spontaneous polarization charge. Variations in the properties of the layers could affect unit-to-unit variability, as well as the overall characteristics of the device.

These issues are representative of the types of questions that arise from the body of literature on compound semiconductors. There are many different types of materials, as well as different ways of using them in devices. Their operation, as well as the relevant material properties, must be understood in order to evaluate the work. Due to the limited market for many compound semiconductors, advanced devices are often available only in small production runs, with limited information about reliability. Reliability is usually evaluated with smaller sample sizes and more limited conditions than for mainstream silicon devices, with the important exception of the cellular telephone industry, which uses large numbers of compound semiconductors.

Highly advanced compound semiconductor devices are used less frequently in space applications. Part of the reason is reluctance by spacecraft designers to use any technology that does not have a record of previous success in space.

GaAs MESFETs are widely used, but they represent an older technology, without the complications of different substrate materials, and have a proven record in many spacecraft. Light-emitting diodes are also widely used, as well as laser diodes.

1.2.2 Radiation damage in light-emitting diodes

In most cases, compound semiconductors have a relatively high tolerance to radiation damage. They are often accepted for use in space without requiring radiation qualification tests on the basis of "similarity" arguments and cost savings. However, reliability continues to be an important concern, and special burn-in and screening tests are usually required, along with reduced power and voltage limits for circuit applications.

While this approach has generally worked well, there are cases where compound semiconductors have failed in space. In at least two instances the failure effectively ended the mission. One occurred on the NASA Galileo mission, a high-profile mission to explore Jupiter and its moons. The radiation belts of Jupiter are more intense than those surrounding the earth. Radiation damage was a key factor in selecting components for this mission.

The spacecraft, launched in 1989, worked for many years, exceeding its design life by about a factor of two. A total of 34 orbits were made around Jupiter during the explorations phase of the mission. However, the tape recorder[*] used for data storage failed during the last orbit. That

---

[*] It may seem strange that a tape recorder was used. However, the components for the mission were selected in the early 1980s when solid-state memories were limited to about 64 kb per chip.

particular orbit made a closer approach to the planet than the previous 33 orbits, which increased the flux of radiation. The failure was traced to a GaAs LED within the tape recorder electronics package. Operation was temporarily restored afterwards (the annealing approach used for recovery is discussed in Chapter 11).

Laboratory tests were made on similar LEDs shortly after the failure occurred in space. Those results, along with the estimated displacement dose during the Galileo mission, are shown in Fig. 1-2 [2]. During the first 33 orbits the LED output power degraded to about 60% of the pre-irradiation value, with no effect on circuit operation. The high radiation level of the last orbit reduced the light output to about 25%, resulting in circuit failure.

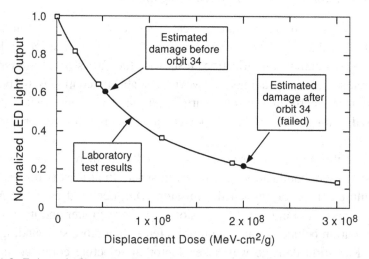

Fig. 1-2. Estimated effect of radiation damage on optical power output from a light-emitting diode in the Galileo spacecraft during the exploration phase around the planet Jupiter.

Despite the high radiation level, this was one of the few electronic or optoelectronic parts that failed during the mission (including many devices fabricated with silicon). A careful series of evaluations and radiation tests had been done on all components used on the spacecraft during the engineering phase.

In the case of the LED, however, the evaluations failed to consider the effects of displacement damage, which causes this particular type of LED to be extremely sensitive to space radiation. We now know that the on-orbit failure could have been avoided by simply using a different type of LED.

This example illustrates the importance of understanding how radiation affects compound semiconductor devices. In retrospect, it is easy to flag this particular LED as a suspect part because of the way in which it constructed. One of the problems with compound semiconductors is the extremely wide range in radiation sensitivity, approximately four orders of magnitude. Although most of them are quite resistant to displacement damage, others are so sensitive that they can fail in space, even in earth-orbiting spacecraft. The key is to understand the device structure and damage mechanisms so that devices that are highly sensitive to radiation damage can be identified.

This brief overview provides some insight into the issues involved for reliability and radiation hardening of compound semiconductors. There are many different types of devices and materials, and only some of the issues relating to them can be discussed in a book of this type. Hopefully the discussion of advanced devices and the underlying physics will allow the book to be useful for many years, even as compound semiconductors continue to advance.

References

1. N. Tsurumi, *et al.*, "GaN Transistors for Power Switching and High Frequency Applications", Digest of Papers from the 2008 Compound Semiconductor and Circuits Symposium, pp. 1–5.
2. G. M. Swift, *et al.*, "In-Flight Annealing of Displacement Damage in GaAs LEDs: A Galileo Story", IEEE Trans. Nucl. Sci., **50**(6), pp. 1991–1997 (2003).

Chapter 2

Semiconductor Fundamentals

This chapter reviews some basic properties of semiconductors, emphasizing aspects that are of particular relevance to compound semiconductors. Two topics that will be important in later chapters are the formation of solid alloy solutions, which allow the electronic and optical properties of a semiconductor to be tailored by varying the alloy composition; and heterostructures, which are formed by dissimilar semiconductors that have different bandgaps. The last section discusses p-n junctions, providing a framework for the discussion of more complex devices in the following chapter.

2.1 Fundamental Concepts

2.1.1 Bandgap

One of the key properties of a semiconductor is the bandgap, defined as the energy difference between the valence band – where electrons are loosely bound to atoms in the host lattice – and the conduction band, where electrons can travel freely within the solid. The bandgap of semiconductors is typically between 0.4 and 4 eV. In an ideal pure semiconductor, electrons are forbidden from occupying the region between the valence and conduction bands.

At very low temperature, all the electrons have low energy, and are confined to the valence band. As the temperature increases, the electrons gain energy, increasing the mean energy as well as producing a distribution of energies among the total number of electrons. The energy distribution of electrons in the valence band can be described by a density of states function, which depends on $(m_{eff} T)^{3/2}$ where m_{eff} is the effective mass, and T is absolute temperature (see [1] and [2] for details). The effective mass can be used to describe many types of interactions of electrons in a solid. It is different from the free electron mass because motion within the crystal is determined by the quantum properties of the crystal lattice. The effective mass depends on crystal orientation, and can even be negative under some conditions.

Several mechanisms, including thermal excitation, can increase the energy of electrons in the valence band. If the electron energy is high enough, an electron can be raised from the valence to the conduction band. The occupancy rate of electrons in the valence band, F(E), can be described by the Fermi-Dirac distribution function

$$F(E) = \frac{1}{1 + e^{(E-E_F)/kT}} \quad (2\text{-}1)$$

where E is the energy of a specific electron, and k is the Boltzmann constant. E_F is the energy at which the probability of occupation of the excited state in the conduction band is ½ (the Fermi level). *The Fermi level can be considered as a reference energy level under equilibrium conditions within a semiconductor.*

For an intrinsic semiconductor (with low impurity density) the Fermi level is located near the center of the bandgap. At very low temperature F(E) is a step function. Figure 2-1 shows F(E) vs. E for three temperatures. In this example, we assume the Fermi level of the material is 0.6 eV. As the temperature increases, the occupancy probability no longer has the sharp dependence that occurs at low temperature, but varies more smoothly with energy for electron energy values that are near E_F.

The temperature sensitivity of F(E) is determined by the quantity E_F/kT; when E_F/kT is $>10^{-3}$ the occupancy rate begins to exhibit a more gradual dependence on electron energy. Semiconductors with large bandgaps require higher temperatures to reach the region where the Fermi–Dirac function begins to become sensitive to temperature. This "smearing" of the edge of the Fermi–Dirac function causes many semiconductor properties to be strongly dependent on temperature.

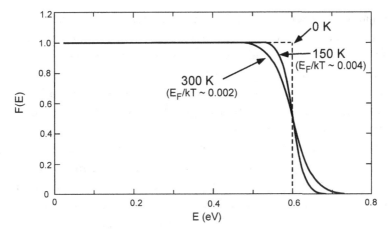

Fig. 2-1. Fermi–Dirac function for the occupancy rate of electrons in the valence band for three different temperatures. The Fermi level in this example is 0.6 eV.

When an electron is elevated to the conduction band it creates a corresponding vacancy (a hole) in the valence band. The holes are effectively particles with positive charge that can move within the valence band in much the same way that energetic electrons move within the conduction band. Electronic properties of semiconductors are often described by the characteristics of electrons and holes. Holes have an effective mass that describes their motion within the valence band, analogous to the effective mass of electrons in the conduction band.

The effective mass of holes is usually larger than that of electrons, due in part to the large number of scattering centers that are present in the valence band.

For metals, the valence and conduction bands overlap, allowing all the outer shell electrons to travel freely within the crystal lattice. Metals have a very high electron density, with ohmic properties that do not have the strong dependence on impurity density and temperature that is exhibited by semiconductors.

For materials with bandgaps between approximately 0.4 and 4 eV the occupancy rate of electrons in the conduction band is high enough at room temperature to make the material weakly conducting. The resistivity depends on basic material properties as well as the density of impurities, discussed below. At room temperature the resistivity of various pure semiconductors lies between approximately 10^{-3} and 10^8 Ω-cm, an extremely wide range. Materials with a bandgap above 4 eV are usually classified as insulators because much higher temperatures are required to reach the condition where significant numbers of electrons begin to occupy the conduction band. Those materials have very high resistivity at room temperature.

The bandgap of a semiconductor also depends on temperature. The temperature sensitivity (at room temperature) is in the range of -0.02 to -0.05% per °C. The temperature dependence is particularly important for optical devices because the wavelength for optical transitions between the valence and conduction bands depends on the bandgap.

2.1.2 Direct and indirect semiconductors

Carrier motion depends on crystal properties as well as the electron occupancy rate described by the Fermi–Dirac function. Particles within a crystal lattice are quantized, with momentum $\hbar k$, where $\hbar$ is Planck's constant divided by 2π, and k is the quantum number for particle momentum within the crystal.

The energy of a particle can be expressed in terms of its momentum by

$$E = \frac{\hbar^2 k^2}{2m_{eff}} \qquad (2-2)$$

Equation 2-2 can be applied to holes as well as electrons, but the effective mass is different. For compound semiconductors the effective mass of holes is much higher than that of electrons, causing the hole velocity to be much lower than that of electrons. This restricts the performance of devices that rely on hole transport rather than electron transport and will be discussed in the next chapter.

The way that recombination occurs when a carrier in the conduction band loses energy, as well as the complementary (absorption) process, is determined by the position of minimum and maximum energy points in the E–k diagram of a crystal. If the minimum energy of the higher level quantum state occurs at the same value of k as the maximum energy level in the lower level quantum state, then it is possible to make a direct band-to-band transition between states by the emission or absorption of a single photon, as shown in Fig. 2-2(a); no change in momentum is required for the transition. However, if the band minima and maxima occur for different values of k, then the transition requires additional momentum (usually from a phonon), resulting in the indirect transition process shown in Fig. 2-2(b).

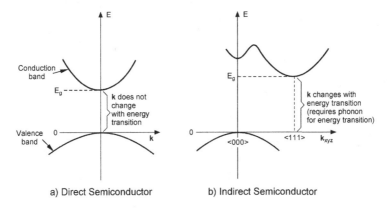

a) Direct Semiconductor   b) Indirect Semiconductor

Fig. 2-2. Direct and indirect transitions in semiconductors.

Note that the momentum from a photon is far too small to provide the amount needed for such transitions, which is the reason that phonon-related processes are required.

For indirect semiconductors, the probability of direct transitions is very low. However, they are often the dominant recombination mechanism for direct semiconductors. For this reason direct semiconductors are usually required for light-emitting and laser diodes.[*] The transition frequency (wavelength) for photons between the two bands depends on the bandgap. Direct transitions involving photons will be discussed in more detail in Chapter 4.

---

[*] Light emission occurs in reverse-biased silicon junctions from impact ionization, as well as in devices with very high impurity densities. However, these processes are many orders of magnitude less efficient than the direct processes that are responsible for photon emission in direct semiconductors.

Some properties of compound semiconductors are shown in Table 2-1, along with silicon. The two entries for SiC correspond to different crystalline structures in that material. InP and GaAs have very high electron mobility, but much lower hole mobility. The low hole mobility restricts the performance of complementary devices for those materials.

Table 2-1. Properties of Some Compound Semiconductors.

Material	Bandgap (300 K)	Band Structure	μ_N cm^2/(V-s)	μ_P cm^2/(V-s)
Si	1.12	Indirect	1350	450
InP	1.35	Direct	4600	150
GaAs	1.42	Direct	8500	400
6H-SiC	3.03	Direct	370	95
4H-SiC	3.26	Direct	720	120
GaN	3.34	Direct	900	180

2.1.3 Carrier densities

2.1.3.1 Intrinsic carrier density

Under equilibrium conditions, thermal energy can cause some electrons to gain enough energy to be raised from the valence to the conduction band. There is a corresponding process where excited electrons will lose energy, and fall back to the valence band. The intrinsic carrier density, n_i, is the density of thermally excited carriers under equilibrium conditions between the two processes. Intrinsic carrier density depends on bandgap as well as temperature, as described by the equation

$$n_i^2 = N_C N_V e^{-E_G/kT} \qquad (2\text{-}3)$$

where E_G is the bandgap, k is the Boltzmann constant, and T is absolute temperature. N_C and N_V are the carrier densities in the conduction and valence bands of the material, which depend on $(m_{eff} T)^{3/2}$. For our purposes it is sufficient to note the general dependence of the carrier densities on temperature and effective mass. Grove [1], and Muller and Kamins [2] provide more details about N_C and N_V.

Figure 2-3 compares the intrinsic carrier density for several semiconductors [3]. Wide-bandgap semiconductors, *e.g.* SiC and GaN, have very low values of n_i which allows them to operate at high temperatures with low leakage current. However, dopant impurities (discussed below) in these two materials are not fully ionized at room temperature, causing some of their properties to be more affected by temperature (at temperatures near room temperature) compared to semiconductors with a smaller bandgap, where all of the impurities are fully ionized at room temperature.

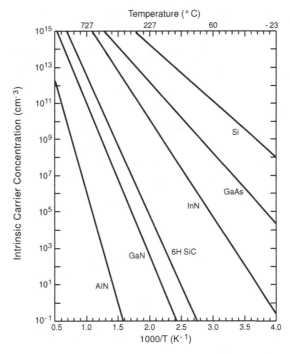

Fig. 2-3. Intrinsic carrier density vs. temperature for various semiconductors [3]. © 2002 IEEE. Reprinted with permission.

2.1.3.2 Doping

Other atomic species can be deliberately introduced into a semiconductor, changing the balance between holes and electrons. These dopant atoms have a different valence than the host semiconductor, and must usually occupy normal lattice sites, not interstitial positions, in order to affect the carrier balance (special steps during processing may be required to activate dopants, moving them to lattice sites rather than interstitial locations).

Impurity atoms with more electrons than the host material act as n-type dopants or donors, providing extra electrons that occupy the conduction band, and increase the electron density. For n-type dopants, the increased electron density in the conduction band leaves positively charged dopant ions in some of the normal lattice sites.

Acceptor atoms have fewer electrons than the host material. They act as p-type dopants, capturing electrons from the electron band, and creating corresponding holes in the valence band.

For either type of dopant, high doping concentrations effectively pin the Fermi level near the energy level of the impurity atoms at room temperature (assuming they are fully ionized). The effect of a high concentration of acceptors on the Fermi level is shown in Fig. 2-4. The acceptor atoms are typically located close to the valence band, and are easily ionized because of the small energy difference between the band edge and the position of the impurity in the energy gap for most materials, even at room temperature (wide bandgap semiconductors are an exception, as well as impurities that have deeper levels within the bandgap).

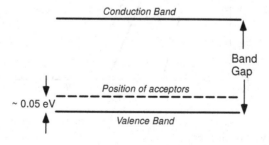

Fig. 2-4. Band diagram showing the position of acceptors near the valence band.

In most cases, nearly all the acceptor ions are ionized at room temperature, and the hole density, p_d, is essentially equal to the density of acceptor atoms, P_d. The minority carrier density, n_n, can then be expressed as

$$n_n \approx \frac{n_i^2}{P_d} \qquad (2\text{-}4)$$

Thus, the doping impurity level determines the minority carrier density. An analogous process occurs for donor impurities, with the donor energy located close to the conduction band, and the hole density determined by the donor concentration.

In some cases donor and acceptor impurities are both present in large concentrations. Conduction in such compensated semiconductors then depends on the net *difference* in doping density. Other semiconductor properties may be affected when this situation occurs, including modification of the band structure because of the large concentration of impurity atoms that are present in the material.

2.1.3.3 Extrinisic and intrinsic conduction

The effect of activated dopant impurities on carrier concentration is fundamental to the operation of most semiconductor devices. Equation 2-4 assumes that the minority carrier concentration is essentially controlled by the majority doping density, which is only valid over a restricted temperature range. Figure 2-5 shows a qualitative picture of the dependence of the Fermi level energy on temperature for a p-doped semiconductor. The dashed line shows the position of the acceptor level within the bandgap.

At very low temperatures none of the impurities are ionized, and the material is effectively an insulator. As the temperature increases the impurities start to become ionized, which defines the extrinsic conduction region. Ionization begins to become important at temperatures on the order of 15–70 K for semiconductors with bandgap energies below 1.2 eV, but the temperature where this transition occurs is higher for materials with a higher bandgap.

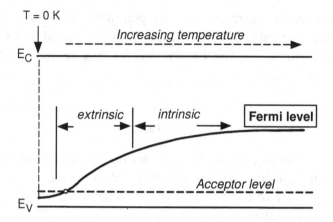

Fig. 2-5. Variation of Fermi level with temperature for a p-type semiconductor showing extrinsic and intrinsic conduction regions.

The dot in the figure shows the condition where the Fermi level is equal to half the acceptor energy level in the bandgap, corresponding to $F(E) = 0.5$, where one-half of the acceptors are ionized. Most semiconductor devices are designed to operate in the extrinsic region, where carrier concentrations are fundamentally controlled by dopant atoms, and the Fermi level is near the acceptor (or donor) level.

From Eq. 2-3, the intrinsic carrier concentration also increases with temperature. We eventually reach the point where, due to the temperature increase, the minority carrier density is comparable to the doping density. When that occurs, the dopant atoms no longer control the carrier density and we reach the region of intrinsic conduction.

The Fermi level, which was approximately equal to the energy level of the dopant atoms in the extrinsic region, increases to roughly the center of the bandgap when we reach the intrinsic conduction region. For silicon, intrinsic conduction occurs at about 200°C, essentially restricting silicon-based semiconductors to temperatures below that temperature. The temperature for intrinsic conduction is much higher – up to 500°C – for wide-bandgap semiconductors (Fig. 2-3 compares n_i^2 for various semiconductors).

2.1.4 Carrier recombination

Several mechanisms can cause recombination of minority carriers in semiconductors. Impurities and crystal imperfections produce localized states that are located within the normally forbidden region of the bandgap. These states can capture minority and majority carriers. Recombination and generation associated with those defects can be described by Shockley-Read-Hall (SRH) recombination, which takes the balance between competing trapping and de-trapping effects into account. We will not discuss SRH recombination here, but refer the reader to works by Sze [4] and Shur [5].

Recombination can also take place at surfaces. Surface recombination depends on the areal density of defects. Several mechanisms can cause surface defects, including the transition from the regular lattice structure of the bulk material towards the edges (boundaries), or at interfaces to other materials, where differences in the internal crystal structure can create dangling bonds. *Surface recombination velocity* is often used to describe surface recombination because the proportionality constant between the carrier density and the recombination rate at the surface has the dimensions of cm/s.

The density of surface states in high quality material is $\sim 10^{10}$ cm^{-2}. Most compound semiconductors have much higher surface-state densities. Special heterostructures (discussed in the next section) are often added to compound semiconductor devices in order to isolate the high surface-state density from the active device region, reducing the effect of high surface recombination rates on device operation.

Band-to-band recombination (which results in absorption or emission of a photon as noted in 2.1.1) and Auger recombination are particularly important for optoelectronic devices, and are discussed in Chapter 4.

2.1.5 Background potential

The energy deep within a semiconductor, *i.e.*, at a distance sufficiently far from regions with other dopants or surfaces to minimize their effects, can be described as a background potential. The background potential for a p-doped material, ψ_F, is related to the intrinsic carrier density and doping level by

$$\psi_F = \left(\frac{kT}{q}\right) \ln\left(\frac{P_d}{n_i}\right) \qquad (2\text{-}5)$$

where q is the electronic charge. An analogous relation can be used for n-doped material. If the doping density is low, the background potential is near the center of the bandgap. It changes with doping concentration, approaching the band edge for high doping densities. The background potential concept is useful in analyzing the effects of surfaces and contacts, as well as in determining the built-in potential of p-n junctions.

2.1.6 Carrier transport

Carrier transport can take place either through diffusion or drift processes. In the presence of a concentration gradient, dN/dx, current from carrier diffusion can be expressed by

$$J = qD\frac{dN}{dx} \qquad (2\text{-}6)$$

where J is the current density, and D is the diffusion constant for majority carriers. The diffusion constant is related to carrier mobility, μ, through the Einstein relation

$$D = \frac{\mu kT}{q} \qquad (2\text{-}7)$$

In the presence of an electric field, carrier mobility due to drift is defined as

$$\mu = \frac{v}{E} \qquad (2\text{-}8)$$

where μ is the mobility in $cm^2/(V\text{-}s)$, v is the carrier velocity, and E is the electric field. Mobility in semiconductors is not constant, but depends on both the electric field and the doping level. For example, Fig. 2-6 shows the dependence of mobility in GaAs on doping concentration, calculated from an analytical approximation by Yuan [6]. The mobility decreases to very low values for high doping levels. The low-field electron mobility of GaAs is about five times higher than that of silicon, which is a major advantage for high-frequency operation. However, as shown in Fig. 2-6, the hole mobility is very low, which is typical of most compound semiconductors.

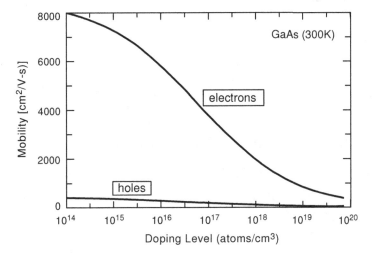

Fig. 2-6. Low-field mobility for GaAs, based on analytical approximations from Yuan [6].

Mobility in all semiconductors decreases with increasing doping concentration due to carrier scattering from impurity atoms. In order to take advantage of the high mobility it is necessary to use very low doping in those regions of a semiconductor device that affect carrier transport.

In indirect semiconductors, carrier velocity increases gradually with increasing electric field, eventually saturating at high electric fields. The saturation drift velocity, v_s, is an important material property, limiting carrier transport at high fields. For silicon, the drift velocity saturates at about 10^7 cm/s for electric fields >4 x 10^4 V/cm. Direct-bandgap semiconductors have a more complicated field dependence. Figure 2-7 shows the field dependence of electron velocity for several compound semiconductors, along with that of silicon [7]. Unlike silicon, at higher fields the electron velocity in many compound semiconductors decreases, although the relative importance of this effect diminishes for high doping densities. This is referred to as velocity overshoot.

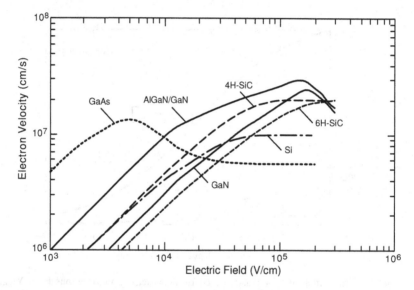

Fig. 2-7. Dependence of electron velocity on electric field for several semiconductors [7]. © 2002 IEEE. Reprinted with permission.

As shown in the figure, the electron velocity of GaAs reaches a maximum at much lower electric fields than for the other materials. That is an advantage when GaAs is used in devices with low electric fields, but it limits the performance of GaAs devices at high fields.

Velocity overshoot does not occur in GaN except at very high electric fields. The high saturation velocity and high electric field that can be applied to SiC and GaN are major advantages for those two materials compared to the others shown in the figure.

2.2 p-n Junctions

2.2.1 Homojunctions

A p-n junction is created when a p-doped semiconductor is placed in contact with an n-doped semiconductor. In a homojunction the two regions are the same semiconductor type, with the same bandgap. The only difference is doping. When the two regions are in contact, minority carrier concentrations are rearranged in both regions near the interface between them. This rearrangement extends over a transition region on either side of the boundary. Within the transition region, minority carrier concentrations are no longer determined by the majority carrier concentration (see Eq. 2-4) because of the influence of charge from dopant atoms on the other side of the interface. We can understand this by considering the Fermi levels in each region.

For an isolated extrinsic semiconductor, the Fermi level is "pinned" near the energy levels of the dopant atoms (see Fig. 2-5). When we consider the two regions in a p-n junction separately they have different Fermi levels, one near the acceptor level and the other near the donor level. However, when they are placed in contact, the Fermi levels in the two regions must have the same value, a condition required for equilibrium; carriers on either side of the junction must have the same energy. As the position of the Fermi level within the bandgap is different for the two doped regions, a step is created in the energy levels of the valence and conduction bands when the two regions are placed in contact. This is shown in Fig. 2-8 for an abrupt n-p junction.

Within the transition region, immobile charge from the dopant atoms is no longer balanced by minority carriers, as shown in Fig. 2-9. In order to maintain overall charge balance, *majority* carriers from each region are injected into the opposite region. The injected majority carriers balance the charge from the immobile dopant atoms at the opposite side of the interface. The electric field within the transition region is also shown in Fig. 2-9.

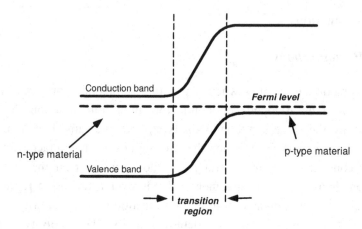

Fig. 2-8. Fermi level of a n-p junction showing the transition region that occurs when two materials with different dopants are in contact.

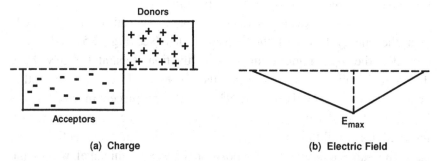

Fig. 2-9. Idealized picture of excess charge and electric field within the depletion region of a step junction.

The transition region is referred to as the depletion region because the minority carrier density is depressed below the level that exists outside the depletion region. Minority carriers are still present within the depletion region, but their concentrations are low enough so that overall charge balance within the depletion region can be approximated by considering only the dopant atoms with that region. The difference in the energy levels of the valence and conduction bands at the p-n junction boundaries creates a barrier for conduction, resulting in highly nonlinear conduction properties.

The width of the depletion layer depends on the number of dopant atoms as well as the material properties. For a step junction under reverse bias, the depletion width, W, is

$$W = \sqrt{\frac{2\varepsilon(N_A + N_D)(V_b \pm V_{app})}{q(N_A N_D)}} \qquad (2\text{-}9)$$

where ε is the permittivity, N_A is the acceptor dopant density on one side of the junction, N_D is the donor dopant density on the other side, V_b is the built-in potential of the material, and V_{app} is the external voltage applied to the junction. The built-in potential either adds or subtracts from the applied voltage.

The built-in potential depends on material properties as well as the doping levels on both sides of the junction, as shown below:

$$V_b = \frac{kT}{q} \ln\left(\frac{N_A N_D}{n_i^2}\right) \qquad (2\text{-}10)$$

As discussed earlier, the intrinsic carrier density, n_i, depends on bandgap (see Eq. 2-3 and Fig. 2-3). Consequently V_b is larger for semiconductors with a high bandgap. Although it varies somewhat with doping level, it is approximately 60% of the bandgap energy for a typical p-n junction. Thus, V_b is about 0.66 V for silicon, 0.85 V for GaAs, and 1.7 V for SiC.

As a result of the offset in the valence and conduction bands, conduction is very different under forward and reverse bias. When the junction is reverse biased, only a small leakage current flows. Under forward bias, the applied potential can overcome the offset in the band structure. Forward current can be described by

$$I = I_s \exp\left[\frac{qV_F}{nkT}\right] - 1 \qquad (2\text{-}11)$$

where I is the forward current, I_s is the saturation current, V_F is the forward voltage, k is the Boltzmann constant, T is absolute temperature, and n is a coefficient called the ideality factor. The forward bias condition where significant current starts to flow is related to the built-in voltage of Eq. 2-10, and is higher for semiconductors with larger bandgaps. Although the diode current depends exponentially on V_F, V_b can be used as a first-order estimate for the forward voltage condition for a diode.

Figure 2-10 shows the dependence of current on voltage for a GaAs diode. Under low forward bias, recombination in the junction depletion region – which competes with the diffusion process that is involved in current flow – decreases the slope of the exponential relationship. The ideality factor is 2 in that region. As the forward current increases, the relative importance of depletion region recombination drops, and diffusion current dominates, with n = 1. The ideality factor, n, is a useful diagnostic parameter for diodes (including light-emitting diodes), as well as for bipolar transistors, and is often used to evaluate the effects of stress or radiation damage.

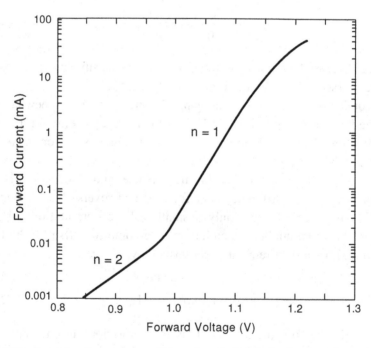

Fig. 2-10. Current–voltage characteristics of a GaAs diode showing the transition between the region dominated by recombination in the space-charge region and the region where diffusion controls the forward current.

For the GaAs diode in Fig. 2-10, a current of 1 mA occurs at a forward voltage of about 1.1 V, about 75% of the GaAs bandgap. I-V curves for a silicon diode are similar to that of Fig. 2-10, but shifted to the left. The nominal turn-on voltage of a silicon diode is 0.7 V. The turn-on voltage of semiconductors with larger bandgaps is correspondingly larger; *e.g.*, about 2.3 V for 6-H SiC. Measurements of the I-V characteristics can be used to determine the approximate bandgap of the material used for diodes, a useful diagnostic method.

2.2.2 Heterojunctions

A junction can also be formed when two semiconductors with different bandgaps are placed in contact, forming a heterojunction (or a heterostructure). In this case the bandgap difference is a result of different material types, not the doping concentration, although doping also influences heterojunction properties. The semiconductors forming the heterojunction can be of different types, or the same basic type with different alloy composition (the bandgap depends on composition).

In a heterostructure the difference in the bandgaps – the band offset – of the two materials introduces a discontinuity in the band diagram that affects the flow of carriers across the junction. Figure 2-11 shows simplified band diagrams for p-n, n-p, p-p and n-n heterostructures. The diagrams correspond to structures where there are abrupt transitions between the different materials. For more gradual interfaces, the discontinuity is spread over a wider region, with reduced step height. Heterostructures of the same type are referred to as isotonic, while n-p or p-n heterostuctures are heterotonic. Another convention is to use an upper case letter for the region with higher doping density, *e.g.* N-n.

Heterostructures are used for several different purposes in devices. Isotonic heterostructures can be interposed between other regions to allow ohmic contacts to be formed, or to serve as barriers between active regions and surfaces. This reduces the effect of surface recombination on device properties because of the band discontinuity. As we will see in the next chapter, very complex device structures can be fabricated with several different heterostructures, using deposition methods such as metal-oxide chemical vapor deposition (MOCVD) to form thin layers with precise thickness and composition.

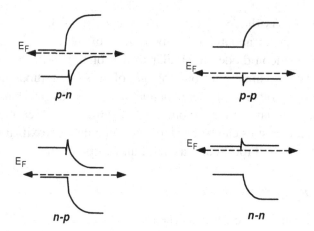

Fig. 2-11. Band diagram of basic heterostructures. Heterotonic structures are at the left, while isotonic structures are at the right of the figure.

Heterotonic structures can be used in much the same way as a basic p-n junction, using the difference in the bandgap of the two materials to increase the effective barrier height compared to a conventional p-n junction. One of the effects of the barrier height is to increase the injection efficiency from highly doped to lightly doped materials compared to injection from a conventional homojunction.

Optical properties of heterostructures are also important, and can be tailored over a relatively broad range. The index of refraction of the two materials forming a heterostructure is not the same. This property of isotonic heterostructures can be used to confine photons within an optoelectronic structure, as well as for other purposes in optoelectronic devices. This is further discussed in Chapter 4.

2.2.2.1 Band offset

The difference in the bandgap of the two semiconductors forming a heterojunction is asymmetrically distributed between the valence and conduction bands. A larger fraction of the band offset usually occurs in the conduction band. For AlGaAs/GaAs heterostructures the conduction band discontinuity is about 60% of the energy gap difference [8]. Similar offset values were obtained for InGaAs/GaAs [9].

Semiconductor Fundamentals   29

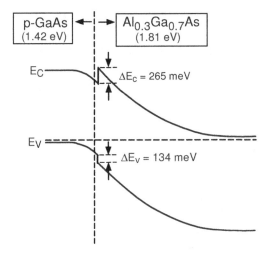

Fig. 2-12. Band diagram of a P-n heterostructure showing how the band offset is distributed between the conduction and valence bands.

Figure 2-12 shows the band diagram of a GaAs/AlGaAs P-n heterojunction. The band offset in this example is 0.39 V, distributed between the valence and conduction bands as shown.

The step in the conduction band acts as an additional barrier for minority carriers, increasing injection efficiency.

Conduction across the heterostructure barrier depends exponentially on the barrier height. The band offset must be large enough to maintain low leakage. This is an important consideration for semiconductors such as InAlGaP, which have small bandgaps, with correspondingly smaller band offsets.

2.2.2.2 Tailoring the bandgap: bandgap engineering

For a heterostructure formed by alloys of the same basic materials, the bandgap is determined by the mole fraction of the semiconductors that comprise the semiconductor alloy. For example, Fig. 2-13 shows the dependence of the bandgap of InGaAs on the mole fraction of gallium, X The bandgap of that ternary material varies over an extremely wide range, depending on composition. A similar figure for AlGaAs is discussed in Chapter 4.

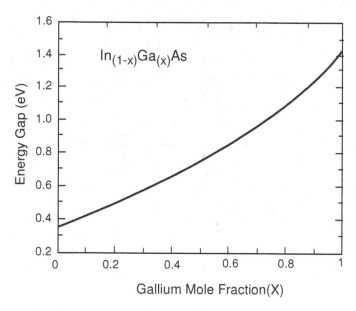

Fig. 2-13. Bandgap of InGaAs vs. the mole fraction of gallium.

The bandgap and the band offset are both dependant on composition. The valence band offset, ΔE_v, of $Al_xGa_{(1-x)}As$ (referenced to the bandgap of GaAs) obeys the equation [10]

$$\Delta E_v = 0.55\, X_{Al} \qquad (2\text{-}12)$$

with X the mole fraction of aluminum. An offset also occurs in the conduction band, but the relationship is more complex, changing slope for X = 0.45, the region where AlGaAs changes from a direct to an indirect semiconductor. The key point is that altering the material composition affects the bandgap, as well as the way that the band offset is distributed between the valence and conduction bands.

Band offset depends on the specific material. For example, the bandgap of pseudomorphic InGaAs/GaAs (where the lattice mismatch produces a strain within the lattice) decreases with increasing indium concentration, but the conduction band offset remains at 60% of the bandgap for all indium compositions [9].

These results illustrate some of the material parameters that are important in fabricating compound semiconductors. The ability to form different regions with precisely controlled thickness, tailored bandgap, and refractive index provides a great deal of flexibility for device design, but requires a thorough understanding of the material properties.

Composition also affects the lattice constant of solid solutions. There are specific values of compositional range for the various materials that provide close lattice matching, and those values are generally used in order to reduce strain and the subsequent crystal defects that occur when strain is present. Some specific lattice-matched values for ternary compounds are listed in Table 2-2. AlGaAs is lattice matched to GaAs over the entire compositional range, but the other materials are only lattice matched for the specific compositions shown in the table.

Table 2-2. Properties of Three Ternary Materials at Lattice-Matched Composition.

MATERIAL	BANDGAP	SUBSTRATE	ΔE_G
$Al_xGa_{1-x}As$	$1.42 + 1.25X$	GaAs	$1.25X$
$In_{0.53}Ga_{0.47}As$	0.81	InP	0.61
$In_{0.52}Al_{0.48}As$	1.51	InP	0.88

It is not always necessary to use lattice-matched materials. Defects from strain can be suppressed if the layer with mismatched lattice constants is sufficiently thin, and the mismatch is not too large. The internal stress is distributed over several lattice sites, altering their location without introducing defects.

The strained region is stable if it does not exceed the critical thickness for the specific material. Figure 2-14 shows the critical thickness for $In_xGa_{1-x}As$ as a function of the indium concentration, X, along with experimental points showing values that have been used to fabricate semiconductor lasers [11]. The critical thickness was calculated from earlier theoretical work [12]. As a rule of thumb, strain from lattice mismatch can be accommodated within regions that are up to 100 Å thick.

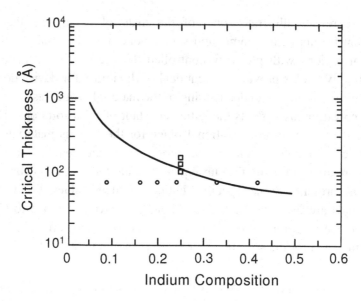

Fig. 2-14. Critical thickness for $In_xGa_{1-x}As$ as a function of indium composition [11]. © 1990 IEEE. Reprinted with permission.

Strained lattices allow other material systems to be used, such as InGaAs/GaAs, where lattice matching is not possible. They also extend the compositional range for adjusting the bandgap beyond the constraints of lattice matching. Other material properties are also affected by strain, such as hole mobility, that can improve device performance compared to lattice-matched materials if the strain is in the appropriate direction.

2.2.2.3 Schottky barriers

Schottky barriers, formed between metals and semiconductors, are frequently used in compound semiconductors, including the gate region of advanced MESFETs (discussed in Chapter 3). The metal region acts as a sink for carriers, and is equivalent to a semiconductor with extremely high doping and negligible thickness. One of the advantages of a Schottky barrier is that it only requires deposition of a thin metallic layer, which is much simpler to fabricate than a diffused (or grown) region with high carrier concentration. The Schottky barrier height depends on a property called the electronic affinity. This is the difference between the background potential and the Fermi level of the semiconductor, although calculations of barrier height from electron affinity do not always agree with experimental results.

Somewhat remarkably, the Fermi level formed between a metal and the semiconductors that we are considering is a constant fraction of the bandgap. Mead and Spitzer [13] showed that the barrier height, Φ_{Bn}, was

$$\Phi_{Bn} = \frac{2E_G}{3} \qquad (2\text{-}13)$$

where E_G is the bandgap. Slight adjustments are required for InAs and InP because the first-order perturbation term associated with the band structure is larger than for other materials. There are also slight differences between different metals, as well as etching methods used in manufacturing, but the relationship of Eq. 2-13 provides a first-order estimate of the Schottky barrier height for most materials.

Schottky diodes do not have the diffusion capacitance of a conventional semiconductor diode, only the depletion capacitance associated with the semiconductor region. This increases the frequency response, allowing them to be used at microwave frequencies. The smaller capacitance is also an advantage when Schottky diodes are used as the gate region of a MESFET or MODFET (see Chapter 3).

2.3 Fabrication and Material Issues

Silicon has several intrinsic advantages over compound semiconductors, along with more than 50 years of development that have improved material properties, and is the material of choice for many different types of electronic devices. One of its main advantages is the ability to grow extremely pure material with very low dislocation densities in large-diameter crystals. This allows very large wafer diameters to be used, thus reducing manufacturing cost. In addition, silicon dioxide is a very high quality insulator that can be easily grown on silicon during fabrication. Comparable insulators are unavailable for most compound semiconductors.

With the exception of Si:Ge (which can be considered as an extension of silicon technology), all compound semiconductors have much higher dislocation densities, increasing the difficulty of fabricating useful devices. Some compound semiconductors require a different substrate material compared to the active device material used for devices because of high material cost or unacceptable numbers of defects in the primary material.

Other advantages of silicon include high hole mobility (about half that of electron mobility), and high thermal conductivity. The maturity and low manufacturing cost of silicon is likely to continue to make it the material of choice, except for specific applications where compound semiconductor materials offer clear advantages. Nevertheless, the market for compound semiconductors continues to increase, particularly for devices used at high frequency, high voltage, and optoelectronics.

We shall briefly review material issues and problems related to manufacturing for several compound semiconductors in the following subsections. In addition to their basic material properties, it is also important to note whether heterostructures can be formed, as well as any unusual material characteristics (such as polarization effects for GaN) that affect device design.

2.3.1 GaAs

Gallium arsenide is the most mature compound semiconductor material. As noted earlier, the high electron mobility of GaAs is an important advantage. However, severe velocity overshoot occurs at moderate electric fields, reducing the performance of GaAs devices at high fields. AlGaAs is lattice matched to GaAs over the entire compositional range, allowing heterostructures to be formed with low dislocation density. Bulk GaAs crystals can be grown with acceptable defect density, allowing fabrication on GaAs substrates. Undoped GaAs substrates have higher resistivity compared to silicon substrates because of the higher bandgap. This is an advantage for microwave devices because it reduces RF losses in the substrate.

2.3.2 InP

Indium phosphide is used as the substrate material for several different heterojunction-based technologies. Phosphorous is relatively unstable, restricting the ability to grow additional layers with phosphide-based compounds on the initial substrate. Therefore most InP-based devices use InGaAs or InAlAs heterostructures for the actual layers that are involved in device operation, using molecular beam epitaxy or MOCVD processing that restricts the temperature rise in the substrate during subsequent processing steps. Those materials have even higher mobility than GaAs, which is an advantage. However, InP is fragile, restricting the wafer size and affecting yield.

2.3.3 SiC

Silicon carbide is an unusual material with a hexagonal structure that results in numerous "polytypes", which occur because of different ways of stacking the hexagonal crystal planes. There are more than 100 different SiC polytypes.

The various polytypes have different physical and electrical properties (including differences in bandgap, as noted earlier). Three polytypes are commonly used: 6H, 4H and 3C. The 4H polytype has higher mobility, making it the preferred material for many device applications. SiC typically has very high dislocation densities, including micropipes that extend over a large range within the crystal, but it can be grown in bulk form. Recently it was announced that 4H material can be produced without micropipes, a major breakthrough for that material [14].

High quality oxides can be formed on SiC. This allows SiC MOS transistors to be fabricated, which is not possible for most other compound semiconductors. However, there is no suitable heterostructure material for SiC. Junctions formed in SiC are usually formed using conventional diffusion methods. The high breakdown field of SiC is a major advantage for high-voltage devices.

2.3.4 GaN

Gallium nitride has only become a viable material within the last 15 years. There are several problems with this material. It is extremely difficult and costly to grow large GaN crystals, and for that reason nearly all GaN devices are formed on other substrates. Silicon carbide and sapphire are both used, but neither material is closely lattice matched to GaN. SiC substrates have higher thermal conductivity compared to sapphire substrates, an advantage for power devices.

Heterojunctions using AlGaN can be formed with GaN. The large energy gap difference is an advantage for devices design. However, spontaneous polarization is present at the AlGaN/GaN interface, along with strain from lattice mismatch. The understanding of those effects has only recently reached the level that is necessary to design reliable devices with GaN. GaN has a high breakdown field and can also operate at high temperatures, allowing microwave devices to be fabricated with very high power densities.

2.4 Power Applications

Applications for devices at high power and high frequency must consider tradeoffs with other material properties. There are many different requirements for power devices, including applications that require efficient performance at low voltage (*e.g.*, cellular phones), microwave applications where more efficient overall operation requires voltages in the 30–80 V range, and very high-power applications where operation at still higher voltages is required for optimized design.

Important material properties for power devices include carrier mobility, maximum electric field (note that velocity overshoot is a limiting factor here), dielectric strength, and thermal conductivity. The critical breakdown field, E_c, is one of the limiting factors. The critical field increases with energy gap, with a nonlinear dependence. Chow and Tyagi [15] used a power law to relate the breakdown field to the bandgap, as shown in Fig. 2-15. The dependence is nearly quadratic with the fit shown in the figure, demonstrating the theoretical advantage of wide bandgap semiconductors for high power applications.

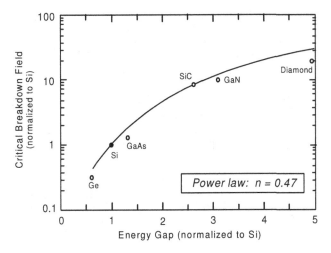

Fig. 2-15. Critical breakdown field vs. energy gap for various semiconductors [15]. © 1994 IEEE. Reprinted with permission.

Various figures of merit have been developed that relate material properties to general classes of power applications. Interpreting various figures of merit (FOM) that have been developed for applications of power devices can be very confusing. Initial work by Johnson [16] used the product of E_c and mobility as a figure of merit. Baliga proposed a cubic dependence on E_c that was revised in a later paper to include switching behavior, resulting in a different dependence on material properties [17]. Huang summarized the earlier work and arrived at a new FOM that was proportional to $E_c\sqrt{\mu}$ [18].

The Huang values are more realistic than the others, reducing the relative advantage of the wide-bandgap semiconductors somewhat compared to the earlier figures of merit. Table 2-3 summarizes those results, along with basic material parameters for several semiconductors. Note that diamond (which we have not discussed here) has the highest overall figure of merit.

An additional concern for power-switching applications is the resistance of the lightly doped drift regions that are needed to withstand high voltage, which is not included in the figure of merit.

Table 2-3. Figures of Merit for Power Applications for Various Semiconductors.

MATERIAL	BANDGAP @ 300 K (EV)	ELECTRON MOBILITY [CM^2/(V-S)]	CRITICAL FIELD (MV/CM)	JOHNSON FOM	BALIGA FOM	HUANG FOM
SI	1.1	1400	0.3	1	1	1
GAAS	1.43	8500	0.4	11	16	3.3
SIC (6H)	2.86	330	2.4	260	13	3.9
SIC (4H)	3.26	700	2.0	410	34	7.5
GAN	3.45	900	3.3	790	100	8.0
DIAMOND	5.95	2200	5.6	5330	1080	23.8

Below 100 V, the resistance of unipolar silicon devices is low enough to meet most applications. At higher voltages compound semiconductors provide a significant advantage. If we compare SiC to silicon, 6H SiC provides a 100X reduction in on resistance, while 4H SiC provides a 2000X reduction [19].

2.5 Summary

This chapter has discussed basic properties of semiconductors and p-n junctions, emphasizing points that are of particular relevance to compound semiconductors. More thorough discussions of semiconductor physics are included in the books by Sze and Shur [4,5], and the reader is encouraged to consult those works for more detailed information on device physics.

There are important differences in the properties of compound semiconductors that determine how they can be applied to specific semiconductor devices. Electron mobility is very high for some materials, but velocity overshoot, which does not occur in silicon, may restrict their applications to low electric field. The low hole mobility of most compound semiconductors is also an important limitation. Nearly all compound semiconductor devices rely on high electron mobility. Complementary structures that depend on hole mobility are rarely used, because the low mobility limits the performance of the complementary device structure.

Semiconductor alloys provide the ability to adjust the bandgap of various material systems over an extended range. Bandgap engineering, in conjunction with heterostructures, provides a tremendous degree of flexibility for transistors and optical devices. Advanced fabrication methods allow layers of different materials to be grown with extremely precise thickness. Many different layers can be used to take maximum advantage of heterostructure properties, optimizing performance and reducing surface recombination losses.

GaAs and other semiconductors with intermediate bandgap are relatively mature, and are used in many applications, including high-performance high-frequency amplifiers. They are also used in a variety of optoelectronic applications, an area where silicon cannot compete because of the indirect bandgap.

Wide-bandgap semiconductors have been used to fabricate many types of experimental devices, demonstrating their potential advantage, particularly for power semiconductors because of the high breakdown field and the ability to operate at higher temperature. Commercial devices are available that use those materials for high-power and high-voltage applications. Device development is continuing at a rapid pace, using special methods to overcome the effects of the large defect density that is present in those materials and reduce manufacturing costs.

References

1. A. S. Grove, *Physics and Technology of Semiconductor Devices*, John Wiley: New York, 1967 (recently reprinted by Dover Press).
2. R. S. Muller and T. I. Kamins, *Device Electronics for Integrated Circuits*, Second Edition, John Wiley: New York, 1986.
3. R. T. Kemerley, H. B. Wallace and M. N. Yoder, "Impact of Wide Bandgap Microwave Devices on DoD Systems", Proc. of the IEEE, **90**(6), pp. 1059–1064 (2002).
4. S. M. Sze, *Physics of Semiconductor Devices*, John Wiley: New York, 1981.
5. M. Shur, *Physics of Semiconductor Devices*, Prentice-Hall: New Jersey, 1990.

6. J. S. Yuan, *Si, GaAs and InP Heterojunction Bipolar Transistors*, John Wiley: New York (1999).
7. R. J. Trew, "SiC and GaN Transistors – Is There One Winner for Microwave Power Applications?", Proc. IEEE, **90**(6), pp. 1032–1047 (2002).
8. S. Blakemore, "Semiconductor and Other Major Properties of GaAs", J. Appl. Phys., **53**(10), pp. 123–181 (1982).
9. Y. Zou, et al., "Characterization and Determination of the Band-Gap Discontinuity of the $In_xGa_{1-x}As$/GaAs Pseudomorphic Quantum Well", Appl. Phys. Lett., **58**(6), pp. 601–603 (1991).
10. J. Batey and S. L. Wright, "Energy Band Alignment in GaAs:(Al,Ga)As Heterostructures: The Dependence on Alloy Composition", J. Appl. Phys., **59**(1), pp. 200–209 (1966).
11. J. J. Coleman, "Strained Layer Quantum Well Lasers", 1990 IEDM Technical Digest, pp. 6.1.1–6.1.4.
12. J. W. Mathews and A. E. Blakeslee, "Defects in Epitaxial Multlayers", J. Crystal Growth, **27**, pp. 118–125 (1974).
13. C. A. Mead and W. G. Spitzer, "Fermi Level Position at Metal-Semiconductor Interfaces", Phys. Rev. A, **114**(3A), pp. A713–A716 (1964).
14. Y. Li, P. Alexandrov, and J. H. Zhao, "1.88-mΩ-cm^2 1650-V Normally on 4H-SiC TI-VJFET", IEEE Trans. Elect. Dev., **55**(8), pp. 1880–1886 (2008).
15. T. P. Chow and R. Tyagi, "Wide Bandgap Compound Semiconductors for Superior High-Voltage Unipolar Power Devices", IEEE Trans. Elect. Dev., **41**(8), pp. 1481–1483 (1994).
16. E. O. Johnson, "Physical Limitations on Frequency and Power Parameters of Transistor", RCA Rev., **26**, pp. 163–177 (1965).
17. B. J. Baliga, "Power Semiconductor Device Figure of Merit for High-Frequency Applications", IEEE Elect. Dev. Lett., **10**(10), pp. 455–457 (1989).
18. A. Q. Huang, "New Unipolar Switching Power Device Figures of Merit", IEEE Elect. Dev. Lett., **25**(5), pp. 298–301 (2004).
19. B. J. Baliga, "The Future of Power Semiconductor Device Technology", Proc. IEEE, **89**(6), pp. 822–832 (2002).

Chapter 3

Transistor Technologies

This chapter discusses various types of transistors that can be fabricated with compound semiconductors. Metal-oxide transistors (MOSFETs), which dominate the commercial semiconductor industry, are less important for compound semiconductors because of the difficulty of growing high-quality insulators for the gate region. Silicon carbide is an exception; SiO_2 can be grown on SiC in much the same way as for silicon, and MOS transistors with potential advantages for high-power applications have been developed using SiC.

We will begin with a brief discussion of junction-field effect transistors (JFETs) – historically the first device based on majority carrier properties, – followed by a discussion of metal-gate semiconductor field-effect transistors (MESFETs), including performance limits imposed by device scaling. The next section discusses modulation-doped transistors (MODFETs) that use heterojunctions for carrier injection, along with quantum confinement. This produces a thin two-dimensional conducting region that can be scaled to smaller dimensions, with higher performance, compared to JFETs and MESFETs. A section on silicon carbide MOSFETs will examine power-device features where the properties of SiC provide key advantages. The final part of the chapter discusses high-performance silicon-based bipolar transistors, which incorporate SiGe, as well as bipolar transistors made with other compound semiconductors.

3.1 Junction Field-Effect Transistors

3.1.1 Basic characteristics

Junction field-effect transistors consist of an epitaxial body region with two diffused contacts, the source and drain. An additional region, the gate, is diffused between the two contacts with the opposite type of doping, forming a p-n junction that penetrates into the body, affecting the conductivity between the source and drain. This depletion region is shown Fig. 3-1. A similar depletion region is also present between the body and back gate (substrate).

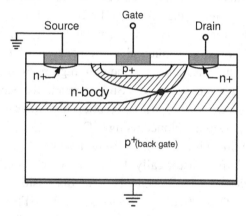

Fig. 3-1. Structure of a basic n-channel junction field effect transistor at pinchoff, when the two depletion regions touch each other. This is the onset of the saturation region.

The presence of the two depletion regions introduces a narrow channel region within the n-body that restricts conduction between the drain and source. Conductivity between the drain and source depends on the way that the depletion regions, formed by voltages at the top and back gates, extend into the body region.

The width of each depletion region depends on the potential between the gate and body. The depletion width increases near the drain because the reverse voltage is higher (the depletion width of a step junction depends on $V^{1/2}$).

The diagram in Fig. 3-1 corresponds to pinch off, the condition where the two depletion regions just touch each other. When this occurs, conductivity between the drain and source no longer depends on drain-source voltage (V_{DS}) because of the overlapping depletion regions (saturation region). When the drain voltage is below pinch off, conductivity between the source and drain depends on drain voltage, increasing with voltage until the pinch-off condition occurs. For low values of V_{DS} the drain current increases linearly with V_{DS}. This is referred to as the linear region, even though the relationship between I_D and V_{DS} becomes nonlinear as V_{DS} increases.

Characteristic curves showing how the drain-source current depends on gate voltage (V_{GS}) are illustrated in Fig. 3-2. The device begins to conduct when the gate voltage exceeds a critical voltage, the threshold voltage, which in this example is about -4.5 V. As the gate voltage is increased, conductivity in the channel region increases, and the current continues to increase with gate voltage until pinch off. The dashed line shows the pinch-off voltage for each value of gate-source voltage. When the drain-source voltage exceeds the pinch off voltage, the drain current no longer depends on drain voltage. However, the pinch-off characteristics of real devices are not as flat as the idealized curves shown in Fig. 3-2.

The gate and body regions are isolated from one another by the junction formed between them. Even under reverse bias a small reverse leakage current will flow between the gate and body region, unlike a MOSFET where the gate current is nearly zero. Higher gate currents occur when the gate is forward biased. For the n-channel JFET shown in Fig. 3-2, gate isolation will no longer occur when the gate voltage exceeds the forward voltage of the gate-source diode (typically ~0.4 to 0.8 V, depending on the work function of the gate and body).

This restricts the range of gate voltages that can be applied in typical circuit applications unless the circuit that drives the gate can provide additional current to the gate region when it is forward biased. Although many conventional circuit applications cannot tolerate the additional loading when the gate is forward biased, RF circuits are often used in that mode.

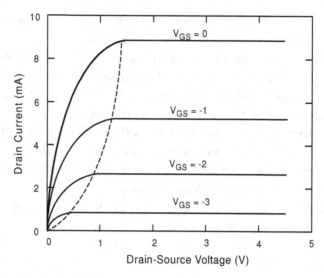

Fig. 3-2. Current-voltage characteristics for an idealized n-channel JFET.

The cutoff frequency, f_T, of a JFET is given by

$$f_T = \frac{g_m}{2\pi(C_G + C_{DG})} \tag{3-1}$$

where g_m is the transconductance, C_G is the gate-source capacitance, and C_{DG} is the drain-gate capacitance. The high mobility of compound semiconductors provides high transconductance, enabling operation at higher frequencies compared to silicon JFETs.

3.1.2 MESFETs

It is also possible to use a metal-semiconductor (Schottky) junction for the gate. This eliminates the diffused p+ region under the gate, making it more compact with lower capacitance and better high frequency performance compared to a JFET. The pinch-off voltage of a MESFET depends on the Schottky barrier height of the metal as well as the doping level in the channel region.

The relationship between pinch-off voltage, V_P, and the properties of the channel is

$$V_P = \frac{q N_D t^2}{2\varepsilon} \tag{3-2}$$

where N_D is the channel doping level, t is the thickness of the channel region, and ε is the permittivity. V_P can be made lower by using thinner channel regions and decreasing the doping level. Depletion-mode (normally on) MESFETs have negative threshold voltages, while enhancement-mode MESFETs have a threshold voltage that is slightly positive. Logic circuits using GaAs MESFETs usually require depletion and enhancement mode devices. P-channel MESFETs can also be fabricated, but their performance is limited by the low hole mobility of most compound semiconductors, and are seldom used.

A basic inverter formed with depletion and enhancement MESFETs is shown in Fig. 3-3. Unlike CMOS inverters, there is a significant load current in the inverter in both states under static conditions (both are n-channel devices, and the load is not in cutoff). The noise margin is also lower compared to CMOS. More complex logic cells are usually used in real circuits, such as buffered FET logic, that incorporate an additional output stage. The output stage in that design requires a negative power supply.

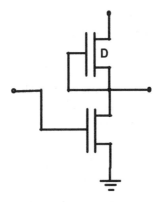

Fig. 3-3. Basic inverter using enhancement and depletion mode MESFETs.

Another logic cell design (Schottky-diode FET logic) uses high-speed Schottky diodes with very small area for logic functions in conjunction with active buffers [1]. It also requires a negative power supply.

Various design methods can be used to improve MESFET performance, including the use of recessed gates to lower gate capacitance. From a general scaling perspective, reducing the gate length increases frequency response, but requires higher channel doping to reduce the depletion width beneath the gate. Theoretical studies of MESFET scaling have predicted that the product ($L^n N_d$) is constant, where L is the channel length, N_d is the channel doping density, and n is an exponent between 0.5 and 2 [2–4].

Golio examined scaling in more detail, and arrived at an empirical result for scaling by comparing published results for functioning devices from several laboratories [5]. He also took the channel mobility at low fields (μ_o) into account, which decreases with increased doping. His work resulted in the equation

$$(\mu_o N_d L) = 6 \times 10^{16} \tag{3-3}$$

Considering the effects of reduced pinch-off voltage from the increased doping concentration, he concluded that scaling in MESFETs was fundamentally limited to gate lengths >0.2 μm. In addition to the effect on V_P, increasing channel doping reduces mobility. Channel doping in a MESFET with 0.5 μm channel length is ~10^{18} cm^{-3}. At that doping level electron mobility in GaAs is less than a third of the mobility in lightly doped GaAs. It falls to even lower values for higher doping concentrations.

Even though MESFET scaling is inherently limited, GaAs MESFETs are widely used in high-frequency power amplifiers. Other materials provide better performance at high voltage, but the maturity, relatively low cost and the established reliability of GaAs MESFETs continues to make them a viable choice for many power applications. Figure 3-4 compares calculations of the power density for MESFETs from GaAs, silicon and SiC based on modeling, along with some experimental comparisons for devices with a channel length of 3 μm [6].

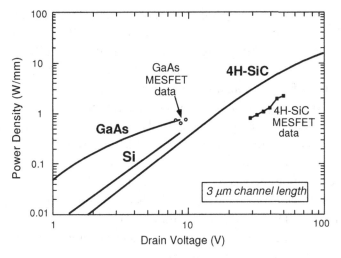

Fig. 3-4. Comparison of power densities for MESFETs from three different technologies with a channel length of 3 μm [6]. © 1995 IEEE. Reprinted with permission.

GaAs performs better at lower voltage, consistent with the high mobility in that material at low electric fields, but its large velocity overshoot reduces this advantage at higher voltages. In contrast, while SiC is clearly a better material at high voltage, it lacks the maturity and established reliability of GaAs.

3.2 Modulation-Doped Field-Effect Transistors

The development of the modulation-doped field-effect transistor (MODFET) was a major advance in device technology, allowing devices to be constructed with higher mobility and smaller gate lengths compared to MESFETs. It is worth noting that heterojunction field-effect transistor (HFET) and high-electron mobility transistor (HEMT), are often used in the literature as an alternative to the term MODFET.

As discussed previously, there is a tradeoff in a MESFET between the doping level in the channel and gate length which limits scaling as well as high-frequency performance. The MODFET eliminates the requirement for high channel doping at short gate lengths, allowing devices to be fabricated with much better frequency response and smaller feature size. Mobility in a MODFET is higher compared to the mobility in a MESFET because the body region of the MODFET is undoped, reducing the effects of impurity scattering. This is a design breakthrough that takes full advantage of the high mobility at low doping concentrations for compound semiconductors.

A MODFET requires a heterojunction between the gate and channel region. The electron and valence bands of the materials forming the heterojunction must line up in such a way that a narrow quantum well is produced within the material with lower bandgap. This forms the conduction region between source and drain [7,8]. Bending of the valence and conducting bands near the heterojunction interface creates the potential well that is required to trap the electrons.

One side of the heterojunction – the "electron supplying region" – is heavily doped, except for a narrow region near the heterojunction interface. The other region, which has a lower bandgap, is undoped. Electrons from the electron supplying region diffuse into the undoped body region, aided by the heterojunction band offset of the two materials as well as the increased potential barrier near the interface that results from band bending. This mechanism can produce large concentrations of electrons in the undoped body region, where they are confined by the quantum well.

A lateral view of an AlGaAs/GaAs MODFET is shown in Fig. 3-5 (*the top of the device is at the left of the diagram*). A gate region, similar to the gate in a MESFET, controls the potential in the doped AlGaAs electron-supplying region. A thin, undoped AlGaAs spacer region is interposed between the heterostructure interface and the GaAs body region where the potential well occurs. The spacer layer has higher bandgap than the GaAs body, and its thickness (sometimes called a setback layer) is an important design parameter.

The device is designed so that only the first two quantum states within the quantum well can be occupied. Because of the quantum well, conduction within the GaAs layer takes place in an extremely thin region – approximately 20 Å – that is effectively two-dimensional and is referred to as the "2DEG" region. The term 2DEG stands for two dimensional electron gas. Note that it is perpendicular to the heterostructure between the AlGaAs and GaAs layers.

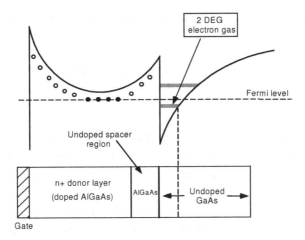

Fig. 3-5. Diagram of the modulation-doped region of a GaAs MODFET where the donor layer is formed with AlGaAs.

A cross-sectional view of the complete device is shown in Fig. 3-6. The spacer layer is the dark region between the undoped GaAs, where the 2DEG layer is created, and the n+ AlGaAs electron-supplying layer. In this example a recessed gate is used to reduce gate capacitance, increasing the frequency response. Doped GaAs "caps" are used to isolate the surface and to aid in forming ohmic contacts.

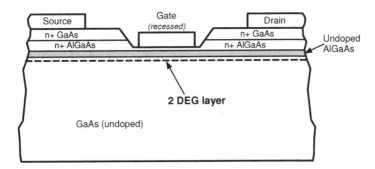

Fig. 3-6. Cross sectional view of a GaAs MODFET with a recessed gate region.

The two-dimensional nature of the conducting region in a MODFET leads to a device model that is based on the sheet charge that is present in the doped AlGaAs layer. If we assume that the sheet charge is proportional to the electric field, the relationship between sheet charge, n_s, and the voltage applied to the gate is given by

$$n_s = \frac{\varepsilon}{q(d_D + d_i + \Delta d)}(V_G - V_{th}) \qquad (3\text{-}4)$$

[9], where d_D is the thickness of the AlGaAs donor layer, d_i is the thickness of the spacer layer, Δd is $\varepsilon a/q$, V_G is the gate voltage and V_{th} is the threshold voltage. The constant a is the (assumed) proportionality constant between the electric field and sheet charge, and ε is the dielectric constant of the undoped region of the heterojunction. Typical values of n_s in AlGaAs are on the order of 10^{12} cm^{-2}.

The linear dependence of n_s on gate voltage in Eq. 3-4 is only an approximation. The relationship becomes nonlinear as the gate voltage is increased. For GaAs the nonlinearity is partly due to unbound (free) electrons that do not contribute to the surface charge density, referred to as DX centers, as well as bound charge in the doped AlGaAs layer that does not contribute to the process involved in creating the 2DEG conducting layer. Figure 3-7 shows the results of a 1-D calculation of the dependence of sheet charge density on gate-source voltage.

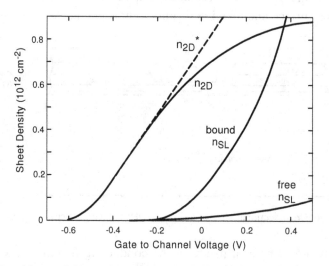

Fig. 3-7. Dependence of sheet charge density on voltage for an AlGaAs/GaAs MODFET [10]. © 1988 IEEE. Reprinted with permission.

The dashed line shows the idealized sheet charge density, while the upper curve shows the gradual decrease in slope for higher gate-to-channel voltages that takes place because of bound and free charge that does not contribute to the sheet charge [10].

Foisy, *et al.* introduced the concept of modulation efficiency to describe the nonlinearity between gate voltage and sheet charge density [10]. In AlGaAs/GaAs MODFETs, the bound charge component of GaAs lowers the modulation efficiency as the gate voltage increases, limiting the sheet charge density to values of about 10^{12} cm^{-2}.

More complicated analytical models have been developed that can be used for detailed modeling of MODFETs [8,11]. Those results show that spacer thickness also affects modulation efficiency. For AlGaAs/GaAs the gain-bandwidth product is about 30% higher for a spacer thickness of 20 Å compared to a spacer thickness of 75 Å.

Initial MODFET development was done using AlGaAs/GaAs where there is close lattice matching between the different materials. It is also possible to fabricate MODFETs with InGaAs, which has a higher sheet charge density because it does not have the DX centers that reduce modulation efficiency at high drain currents in GaAs. However, InGaAs is not lattice matched to either AlGaAs or InAlAs (both materials can be used for the doped region that establishes n_s), resulting in a strained lattice. MODFETs fabricated in this way are referred to as *pseudomorphic*. Figure 3-8 compares the frequency response of a conventional GaAs MODFET with a pseudomorphic InGaAs MODFET [10]. Both devices have channel lengths of 1 µm. The pseudomorphic design has a higher cutoff frequency even in the low current region. The difference in performance is much more significant at higher currents because the modulation efficiency is higher for InGaAs, as previously discussed.

The best performing InGaAs MODFETs have been fabricated on InP substrates, partly because it is possible to use a higher indium concentration. However, InP is fragile and relatively expensive. GaAs substrates can also be used by interposing metamorphic buffers that serve as transition layers, accommodating some of the strain from lattice mismatch in the buffer, reducing the defect density in the active layer.

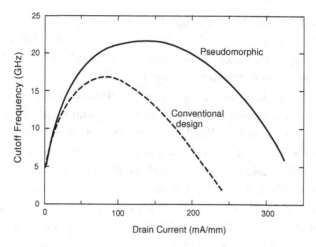

Fig. 3-8. Comparison of the frequency response of a conventional GaAs MODFET with a pseudomorphic MODFET using InGaAs. Both devices have a 1 μm channel length [10]. © 1988 IEEE. Reprinted with permission.

There is a tradeoff in selecting the indium mole fraction. The sheet carrier density is higher for lower InAs concentrations, but mobility is lower. Optimum performance for InGaAs on GaAs occurs for a mole fraction of 0.4 [12], compared to InP substrates which typically use a mole fraction of 0.53. The higher bandgap that results from lower indium concentration also increases the Schottky barrier height, allowing higher gate voltages to be used. Comparable performance has been obtained with both substrate materials.

The properties of the three material systems used for high-speed MODFETs are listed in Table 3-1. The maximum sheet charge density is higher for InAs, partly because of the higher band offset in that material.

Table 3-1. Properties of Material Systems for High-Speed MODFETs.

MATERIAL	SUBSTRATE	BAND OFFSET (EV)	MAXIMUM SHEET CHARGE DENSITY [cm^{-2}]
AlGaAs/GaAs	GaAs	0.25	8×10^{11}
AlGaAs/InGaAs	GaAs	0.45	2×10^{12}
InAlAs/InGaAs	InP	0.60	4×10^{12}

As discussed in the last chapter, the high breakdown field of GaN results in better performance for high-power applications than in other materials. Commercial GaN power MODFETs are now available that use AlGaN/GaN heterostructures, but as discussed in Chapter 2, SiC or sapphire substrates are required because GaN substrates are unavailable (the material is extremely difficult to grow in large crystals). As both substrate materials are poorly lattice matched to GaN, a nucleation layer is typically used between the substrate and the active device layers to absorb some of the strain and reduce the dislocation density. There is also a proprietary process for manufacturing GaN on silicon substrates, which lowers manufacturing cost.

The basic principles of GaN MODFETs are the same as for other material systems. However, a polarization charge is present between the AlGaN used to form heterostructures and GaN that affects that material system. Figure 3-9 shows the bandgap and piezoelectric constant for AlGaN/GaN [13]. InGaN produces a net piezoelectric charge at the InGaN/GaN interface. This piezoelectric charge introduces a surface charge density that is large enough to cause a 2DEG region to form in the heterostructure, even without doping the InGaN electron-supplying layer.

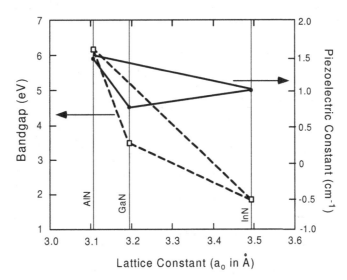

Fig. 3-9. Relationship between lattice constant, bandgap and polarization for the AlGaInN system [13]. © 2002 IEEE. Reprinted with permission.

The effect of the polarization charge needs to be incorporated in device models for GaN MODFETs because it is a significant fraction of the net surface charge. Sheet charge densities up to 5×10^{13} cm^{-2} can be obtained in GaN, much higher than the maximum sheet charge in GaAs. The larger sheet charge density partially offsets the disadvantage of the lower electron mobility in GaN.

GaN MODFETs can exhibit current collapse when they operate at high voltages (this is an important reliability issue that will be discussed in more detail in Chapter 6). Surface passivation can reduce, but does not eliminate current collapse. One way to reduce gate collapse for high voltage devices is to add a field plate that overlaps part of the gate region in the direction towards the drain. The field plate is essentially an extension of the gate, as shown in Fig. 3-10 [14]. AlGaN is used as the gate region, creating a 2DEG region in the GaN layer. The GaN layer is grown by molecular beam epitaxy on a sapphire substrate. A silicon-nitride layer is used to isolate the field plate. This particular structure has a breakdown voltage of 570 V, with an f_T of 6.2 GHz.

Although the additional capacitance of a field plate reduces high-frequency performance, it allows structures to be fabricated with much higher voltage, a major advantage for millimeter wave applications.

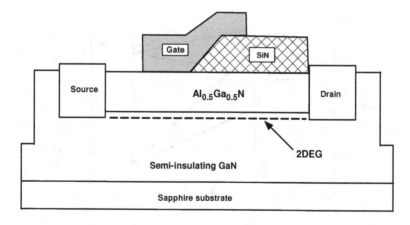

Fig. 3-10. AlGaN/GaN HEMT with a field plate to reduce current collapse [14]. © 2000 IEEE. Reprinted with permission.

Other GaN structures have been developed with larger overlap regions, along with recessed gates that increases transconductance. A MODFET with this design, optimized for voltages below 50 V, produced a saturated power output of 197 W [15]. Although extending the field plate increases feedback capacitance, that was partially compensated by operating the device at higher voltage.

The power density per unit length (W/mm) is often used as a figure of merit for power applications. GaN MODFETs have been designed with power densities less than 10 W/mm, about a factor of ten higher than the power density of GaAs MESFETs or silicon-based technologies. Operating voltage is also an important consideration because the voltage requirements are very different for specific applications. GaAs MESFETs, which are relatively mature, provide high output power for voltages below 15 V, but their performance suffers at higher voltage for various reasons, including velocity overshoot.

GaN MODFETs have been developed for applications up to 30 V, including one technology that uses silicon substrates, reducing manufacturing cost. SiC MESFETs are available for applications at higher voltage.

3.3 MOS Transistors

3.3.1 Fundamental concepts

The operation of a basic MOS transistor is somewhat similar to that of the basic JFET, described in Fig. 3-1. The main difference is in the gate, which for a MOSFET is a high-quality SiO_2 region that has no direct contact to the channel region, and extremely low leakage current (except for highly scaled devices with very thin gates). Control of the current flow between the source and drain is determined by the presence of an inversion layer, formed beneath the gate, which produces a conducting n-channel within the p-well. Figure 3-11 shows a basic diagram of an n-channel MOSFET within an integrated circuit. Shallow SiO_2 trench regions are used for lateral isolation in this structure.

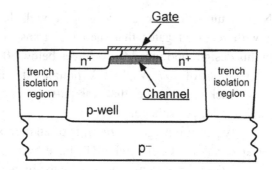

Fig. 3-11. Diagram of a scaled NMOS transistor. Lateral isolation is provided by shallow trench isolation.

The current-voltage characteristics of a MOSFET are similar to those of a JFET, including a pinch-off region at higher drain voltage where the current no longer depends on the gate voltage. A logarithmic plot of the current of an enhancement-mode silicon MOSFET in the linear region is shown in Fig. 3-12. The current increases logarithmically with gate voltage over many decades. The key parameters are the threshold voltage, which is 0.25 V, the saturation current, and the sub-threshold slope. Ideally the sub-threshold slope at room temperature is 60 mV/decade, but other factors cause it to be closer to 80 mV/decade for typical devices.

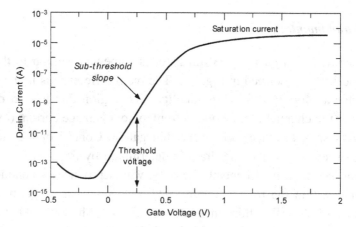

Fig. 3-12. Semi-logarithmic plot of drain current vs. gate voltage showing the threshold voltage, saturation current, and sub-threshold slope.

The threshold voltage is determined by the work functions of the gate and silicon. It can be adjusted by implanting a region with higher doping density in the channel just below the gate. The drain current in saturation can be approximated by the equation

$$I_D = \frac{\mu C_{ox} W (V_G - V_T)^2}{2L} \qquad (3\text{-}5)$$

where µ is the mobility in the channel, W is the channel width, L is the channel length, and C_{ox} is the capacitance per unit area of the gate oxide. The current is proportional to mobility and C_{ox}. The C_{ox} factor is one reason for the improved performance of silicon MOSFETs with scaling.

The electron mobility of 4H-SiC is only about half that of the electron mobility in silicon (see Table 2-1). Thus, SiC MOSFETs are at a disadvantage compared to silicon except when high fields are required, where SiC has a higher electron velocity. However, they also benefit from a lower theoretical value of resistance in the drift region. Early SiC MOSFETs designs suffered from low channel mobility and much higher specific on resistance than expected [16]. These limitations were solved by changing the device structure, using a depleted region between the drift region and the structure at the surface.

Figure 3-13 shows a diagram of a 10-kV power MOSFET [17]. In this design the additional depleted regions produce a JFET that is formed between the drift region and the MOSFET-like region at the surface. Note that current in the MOSFET flows in the lateral direction.

The structure operates as follows: Electrons flow laterally from the n+ source through the channel (in the p-well region) beneath the gate. Once they reach the JFET region, they flow vertically through the JFET to the n- drift region to the drain. The p+ regions adjacent to the source are part of the isolation structure, which is necessary because of the high voltage.

The gate oxide thickness is 60.2 nm, considerably thicker than gate oxide in scaled silicon CMOS, but compatible with the higher gate voltage requirements for power devices.

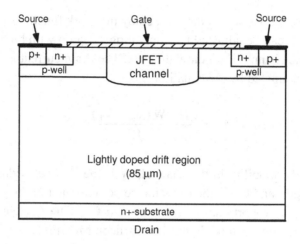

Fig. 3-13. Diagram of a SiC DMOS structure with a voltage rating of 10 kV [17]. © 2004 IEEE. Reprinted with permission.

The mobility, extracted from the saturation characteristics, is only 22 cm^2/V-s for this design, probably because of a high interface trap density. The specific on resistance is 0.12 Ω-cm, which is about 85 times lower than the theoretical value of a silicon device. This illustrates the advantage of SiC MOSFETs for power semiconductors with high voltage ratings. Research is continuing on SiC MOSFETs, and it is likely that there will be additional improvements in the future.

3.4 Bipolar Transistors

3.4.1 Fundamental concepts

A bipolar transistor is formed when two back-to-back p-n junctions are placed in close proximity, separated by a thin base region. A highly simplified diagram of such a structure is shown in Fig. 3-14. Operation of a bipolar transistor depends on injection of majority carriers from the emitter into the base region, (where they are minority carriers) increasing the minority carrier density in the base. If the base region is narrow, then nearly all the carriers injected from the emitter will be transported through the base into the collector region (where they are once again majority carriers).

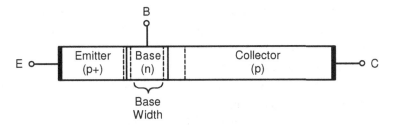

Fig. 3-14. Diagram of an elementary bipolar transistor.

For transistors with wide base regions, recombination within the base limits the gain, although recombination in the space-charge region and surface recombination are also factors. The fractional number of carriers that actually reach the collector is called the base transport efficiency.

In modern HBTs with very thin base regions, reverse injection from the base to the emitter is often the limiting factor for transistor gain, not base transport efficiency. Reverse injection is more important in high-frequency HBTs because higher base doping is required to reduce base resistance, which in turn decreases injection efficiency. Recombination and reverse injection are effectively loss terms, reducing the number of carriers that are injected from the emitter to the collector, requiring additional base current to compensate for the lower carrier density.

The common-emitter current gain, h_{FE}, is one of the key parameters for bipolar transistors. It is defined by

$$h_{FE} = \frac{I_C}{I_B} \qquad (3\text{-}6)$$

where I_C is the collector current, and I_B is the base current. The parameter h_{FE} depends on operating current and the voltage applied between the collector and base terminals. For a conventional transistor h_{FE} increases with collector-emitter voltage.

The voltage dependence is reflected in a related parameter, the Early voltage, which is the extrapolated value of the reciprocal of the proportionality constant between the rate of change of h_{FE} with collector-base voltage and the collector voltage [18]. Although that sounds complicated, it is simply the slope of the current dependence at higher values of collector-emitter voltage extrapolated back to the x-axis.

For a conventional npn transistor the Early voltage is negative. The slope of the voltage dependence is positive, causing the x-axis intercept of the slope to occur at a negative value of collector voltage. A large Early voltage implies a small dependence of gain on voltage, which is an advantage for analog applications where the output resistance of the transistor needs to be very high.

Transistors that are optimized for high-frequency performance often operate near the region where avalanche breakdown occurs, particular for silicon technologies. Figure 3-15 shows the dependence of h_{FE} on voltage for a high-frequency transistor. In this case the transistor design has been modified, using very thin regions with higher doping levels.

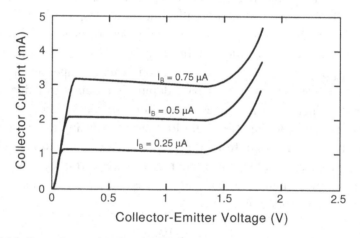

Fig. 3-15. Dependence of collector current on collector-emitter voltage for a high-frequency transistor intended for logic applications.

Consequently the avalanche breakdown region– which can be more than 100 V for conventional, general purpose transistors – is below 2 V for this device. Breakdown voltage is an important limitation for high-frequency devices. The product of the gain-bandwidth and the collector-emitter voltage is sometimes used as a figure of merit for very high frequency transistors.

Another feature of this transistor is that the gain decreases slightly with increasing collector–emitter voltage (until the voltage is high enough to approach avalanche breakdown), which is not the case for most transistors. This results in a positive Early voltage instead of the negative value for conventional transistors.

Although the parameter h_{FE} considers a transistor as a current-controlled device, with the collector current controlled by the much smaller base current, it is also possible to evaluate a transistor from the standpoint of voltage control. Collector and base currents depend on the forward base-emitter voltage, V_{BE}, as follows:

$$I_C = \frac{qD_n n_b}{W_b} \exp\left(\frac{qV_{BE}}{kT}\right) \quad (3\text{-}7)$$

$$I_B = \frac{qD_p p_e}{W_e} \exp\left(\frac{qV_{BE}}{kT}\right) \quad (3\text{-}8)$$

One way of presenting transistor characteristics involves simultaneous plots of base and collector current vs. base-emitter voltage, V_{BE} (Gummel plot). An example is shown in Fig. 3-16. The slope of the collector current is constant over a wide range of forward voltage, only departing from linearity at high currents because of series resistance in the base and collector regions. The slope of the base current is not constant. It increases by a factor of 2 as the injection level reaches the stage where recombination in the emitter-base space charge region competes with transport through the base region.

The gain-bandwidth product (f_T) of a bipolar transistor is determined by fundamental time constants associated with the transit of carriers through the emitter, base and collector, along with the RC time constants that are associated with the emitter and collector. The RC time constants decrease with increasing emitter current.

For older transistors with uniform base doping, f_T at high currents is limited by the base transit time, τ_B

$$\tau_B = \frac{1}{2\pi f_{T\,max}} = \frac{x_B^2}{2D} \quad (3\text{-}9)$$

where x_B is the base width, and D is the minority carrier diffusion constant in the base region. For transistors with graded base doping, this relationship can be modified by integrating the doping profile in the base region (see Ref. 18 for more details).

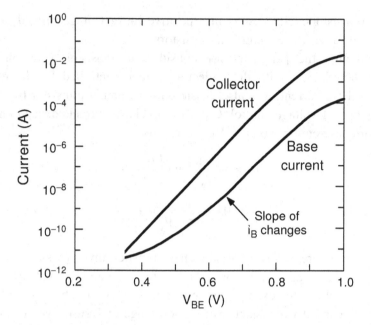

Fig. 3-16. Gummel plot for a silicon HBT transistor.

The value of $f_{T\,max}$ in Eq. 3-9 is the limiting value at very high currents Ignoring the small transit time associated with the emitter, the dependence of f_T on current is given by

$$f_T = \left(\tau_B + \tau_C + \frac{kT}{qI_C}(C_{BE} + C_{BC}) \right)^{-1} \qquad (3\text{-}10)$$

where τ_C is the transit time through the collector, I_C is the collector current, C_{BE} is the emitter-base junction capacitance, and C_{BC} is the base-collector junction capacitance.

The limiting value ($f_{T\,max}$) can be determined by measuring f_T in a circuit with very low collector resistance (this eliminates the effect of external RC time constants, along with the Miller effect, which effectively multiplies the base-collector capacitance by the voltage gain of the circuit). There is a linear relationship between $1/f_T$ and $1/I_C$. The extrapolated Y intercept corresponds to the first two terms in Eq. 3-10.

3.4.2 Modern silicon and silicon-germanium transistors

Conventional bipolar transistors require high emitter doping in order to increase the effectiveness of carrier injection from the emitter to the base. The high doping level suppresses reverse injection from the base to the emitter.

A new approach was developed in the 1980's where doped polysilicon, originally proposed to replace metal contacts, actually functioned as a more efficient emitter into the p-base region than a conventional n+ doped emitter [19,20]. A diagram of test structures that were used in the initial studies is shown in Fig. 3-17 [21]. This breakthrough allowed extremely shallow emitter regions to be used, on the order of 30 nm, decreasing the extrinsic base thickness.

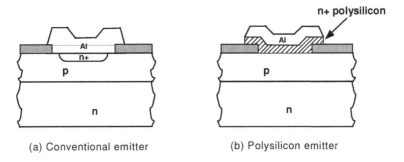

(a) Conventional emitter   (b) Polysilicon emitter

Fig. 3-17. Test structures using conventional and doped polysilicon emitters [21]. © 2001 IEEE. Reprinted with permission.

The polysilicon emitter functions in much the same way as highly doped crystalline silicon, in that holes injected in the reverse direction (from the base to the emitter) recombine within the n+ polysilicon, rather than at the interface between the polysilicon and the p-base material. Narrow-base transistors with polysilicon emitters typically have gains above 400 because of the high charge injection efficiency.

Other modifications to increase the frequency response and reduce device area include self-aligned base contacts; a pedestal collector that provides a top contact to the collector through an extension from the highly doped sub-collector; and advanced isolation methods, such as shallow trench isolation that have lower capacitance compared to conventional junction isolation.

The other breakthrough in silicon HBT technology is the incorporation of SiGe in the base. SiGe has a lower bandgap than silicon, which increases the efficiency of hole transport through the base. A graded concentration is typically used. Although SiGe has a lower bandgap, the trapezoidal concentration dependence produces only a small band offset, as shown in Fig. 3-18 [21].

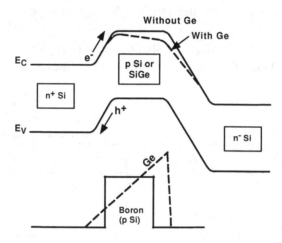

Fig. 3-18. Band diagram of a conventional and SiGe high frequency transistor [21]. © 2001 IEEE. Reprinted with permission.

It is the graded profile rather than the band offset which is responsible for the increase in performance of SiGe transistors. The graded SiGe base increases current gain as well as the Early voltage, while decreasing the transit time.

3.4.3 Other materials

Silicon (extended to incorporate Si-Ge) is still the dominant technology for advanced HBTs that do not require high breakdown voltage because of the maturity of silicon processing, high thermal conductivity, low fabrication cost and high yield.

The most advanced Si-Ge transistors operate at very high current densities, under electric fields that are high enough to approach the avalanche region. Those details, and their impact on reliability, are discussed in Chapter 6.

Despite the dominance of silicon-related processes, higher performance HBTs can be fabricated with other materials, including AlGaAs [22]. Those devices are typically designed with a significant band offset between the emitter and base, which is more efficient for forward carrier injection and reduced reverse injection compared to the graded bandgap structure of SiGe transistors. Figure 3-19 shows the structure of an advanced GaAs HBT. AlGaAs is used for the emitter, and a p-doped AlGaAs region is used for the base, with a different Al fraction than the emitter, to form the emitter-base heterostruture. The collector is GaAs, forming an additional heterostructure for the base-collector junction.

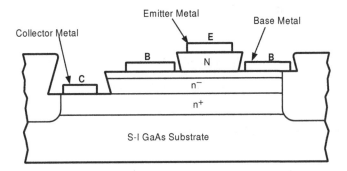

Fig. 3-19. Diagram of a GaAs heterojunction bipolar transistor.

The reduced emitter-base capacitance and higher mobility of GaAs provides somewhat better frequency response at much lower current densities compared to Si-Ge, as shown in Fig. 3-20 [21]. The two structures have comparable emitter areas, and represent processes with comparable feature size.

The higher frequency response at low current allows GaAs HBTs to be scaled to smaller dimensions than SiGe devices with the same feature size. However, this comparison is somewhat academic because it does not consider the low manufacturing cost and very small feature size (0.045 μm in 2008) of silicon-based technologies. Chapter 6 will discuss SiGe transistors in more detail, showing how very high frequency response can only be obtained at very high current densities that can potentially affect reliability.

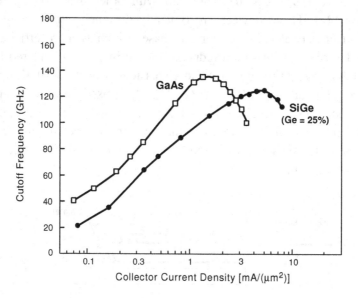

Fig. 3-20. Comparison of GaAs and SiGe HBTs with comparable processing technologies [21]. © 2001 IEEE. Reprinted with permission.

As discussed in the previous chapter, SiC has better properties for high power, high voltage applications, although SiC HBTs have lower gain than silicon-based transistors. Figure 3-21 shows the dependence of f_T on emitter current density for silicon carbide HBTs [23]. If we compare the current density with that of the SiGe and GaAs devices in Fig. 3-20, the maximum f_T for SiC occurs at a current density that is about two orders of magnitude lower than for the other technologies.

There is a factor of 10^5 in converting between current density units in the two figures. Even though the unity gain frequency of the SiC devices is lower, the low current density and high field provide a net performance advantage for high-voltage devices. Another advantage of

SiC is high-temperature operation. SiC can operate at temperatures up to 300°C, allowing higher power densities to be achieved compared to HBTs with lower bandgap semiconductors.

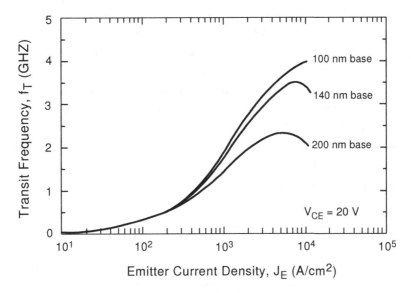

Fig. 3-21. Dependence of f_T on emitter current density for SiC BJTs [23]. © 2005 IEEE. Reprinted with permission.

The relatively low gain of SiC BJTs has been attributed to recombination in the emitter-base space-charge region. Recent advances in SiC BJT technology have been made by using epitaxial emitters instead of ion implantation. Transistors have been fabricated with current gains >40 at room temperature, with collector-emitter breakdown voltage of 1600 V [24]. The gain is affected by temperature, decreasing about 20% at 150°C, and increasing at higher temperature.

The negative temperature coefficient of gain occurs due to the incomplete ionization of base dopant atoms at room temperature. Once the acceptors are fully ionized the gain increases with temperature because the base and emitter lifetimes in SiC also increase with temperature.

Darlington BJTs have also been developed in SiC at the laboratory level, using a hybrid approach. A gain of 430 has been reported at a collector current of 23 A with a 500 V collector–emitter breakdown voltage [25]. The hybrid design has better performance at high currents than monolithic Darlington transistors. The internal losses of Darlington transistors are higher than for a single transistor in switching applications, but the increased gain reduces the base drive, which may provide a net advantage in performance.

The main thrust of SiC BJT development has been for high-voltage switching applications. Several research groups have demonstrated devices with excellent power and breakdown characteristics, but commercial devices are not yet available. SiC BJTs do not exhibit "power slump" that occurs for MODFETs, providing a potential advantage for high-voltage applications.

3.5 Noise

Noise is an important consideration for many applications. Sources of noise include frequency-independent terms associated with power fluctuations in resistors (or their equivalent terms in active device models); and fluctuations in current flow. At low frequencies, noise from traps introduces a "1/f" noise component that increases inversely with frequency. MESFETs (and MODFETs) have inherently lower noise figures at high frequencies compared to HBTs. This is because noise associated with the base resistance and base current causes the noise figure of an HBT to increase quadratically with frequency for an HBT. However, as those terms are absent for MESFETs, the noise figure increases linearly with frequency, providing a distinct advantage.

The noise figure is a fundamental way to specify noise. At high frequencies where the 1/f noise component can be neglected, the noise figure of a network, N_F, is given by

$$N_F = 1 + \frac{N_a}{GkTB} \quad (3\text{-}11)$$

where N_a is the noise that is introduced by the network, G is the network gain, k is the Boltzmann constant, T is absolute temperature, and B is the bandwidth.

Fukui developed a semi-empirical noise model that can be used to determine the minimum noise figure of a MESFET [26]. With that approach, the minimum noise figure is given by

$$N_{F\min} = 1 + 2\pi k_1 f C_{gs} \sqrt{\frac{R_g + R_s}{g_m}} \qquad (3\text{-}12)$$

where $N_{F\min}$ is the minimum noise figure, k_1 is a fitting factor (approximately 2.7), f is frequency, C_{gs} is the gate capacitance, g_m is the transconductance, R_g is the gate resistance, and R_s is the source resistance. The noise figure is more sensitive to C_{gs} than to the other parameters in this equation.

C_{gs} and g_m are related to gate length, L_g. This allows Eq. 3-12 to be rewritten as

$$N_{F\min} = 1 + k_2 f L_g \sqrt{g_m(R_g + R_s)} \qquad (3\text{-}13)$$

where k_2 is 0.27. Equation 3-13 shows that the noise factor scales inversely with gate length. Experimental studies have verified the validity of this dependence for gate lengths as small as 0.05 μm [27]. Transconductance is also affected by gate length, and a more rigorous evaluation results in a more complex equation with several terms. The net effect on gate length is very small, effectively reducing the dependence to $(L_g)^{5/6}$.

The equivalent circuit in Fig. 3-22 shows various terms that contribute to high-frequency noise. In this circuit the drain resistance is divided into two terms, R_{di} which is the effective output resistance of the device, and $R_{d(ext)}$ which is the additional resistance between the edge of the depletion region and the drain contact. The latter term can be reduced by modifying the device geometry. Similarly, the source resistance, R_s, is the sum of two terms, one involving the semiconductor region between the edge of the depletion region and the source, and the other the resistance of the source contact and metallization.

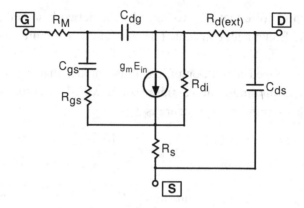

Fig. 3-22. Equivalent circuit of a MESFET or MODFET for noise analysis.

R_M is the resistance of the gate metal stack, which is an important noise source for devices of this type. In this circuit representation the input signal, E_{in}, is the input voltage that is developed internally after ohmic drops through R_M and R_s, not the input signal between the external gate and source terminals.

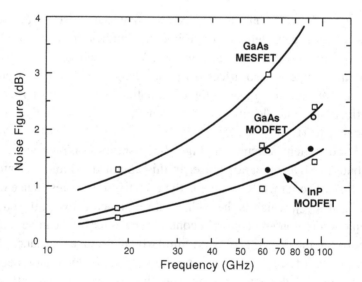

Fig. 3-23. Comparison of high-frequency noise of various device technologies [8]. © 1992 IEEE. Reprinted with permission.

Equations 3-12 and 3-13 show that the lowest noise figure will occur for devices with high transconductance and small gate lengths. From previous device comparisons, we can see that the very high mobility of InP-based MODFETs would provide a lower noise figure, which is indeed the case. Figure 3-23 shows the noise figure vs. frequency for three different device types [8].

The noise figure of the GaAs MODFET is much lower than that of the MESFET because the mobility of highly scaled MESFETs is reduced due to the need for high channel doping.

3.6 Summary

This chapter has reviewed the fundamental operation of JFETs, MESFETs, and MODFETs, all of which can be fabricated with several different types of compound semiconductors. Design issues and performance trends were discussed for devices fabricated with various materials. A brief treatment of SiC MOSFETs was also included because of their importance in high-voltage, high-power applications.

MESFETs and MODFETs using advanced materials provide far better performance at very high frequencies and moderate-to-high voltages than silicon-based technologies. MODFETs can be scaled to smaller dimensions than MESFETs, using a two-dimensional conducting region in undoped material.

The low impurity concentration reduces scattering losses, which maintains high mobility even in devices with very small feature size. MODFET technologies have been developed in several materials, including GaN where the very high critical field provides a major advantage for power applications.

Bipolar transistors were reviewed, followed by a discussion of noise, which is often a discriminating parameter when high-frequency semiconductor devices are used. SiGe technology – which is much more mature because of the similarities in processing technology to conventional silicon devices – has clear advantages for high-speed low voltage applications as well as for integrated circuits that require large numbers of individual components. Other bipolar technologies are required for high-voltage applications, but their development has proceeded more slowly because of the limited market as well as the difficulty of overcoming new technical problems.

Much of the discussion in this chapter has involved new material combinations, with properties that are less well known than for mainstream semiconductors, such as silicon and GaAs. The new materials have much larger defect densities because of basic properties (SiC) or the need to grow the material on substrates with mismatched lattices (*e.g.*, GaN), making it more difficult to develop devices that will meet performance and reliability requirements. It is important to distinguish between potential performance advantages that can be achieved at the research level, and the realities imposed by manufacturing cost, technical difficulties and reliability. In many instances older technologies, such as GaAs MESFETs, continue to be used because the manufacturing challenges have been solved, with demonstrated reliability in critical field applications.

References

1. S. I. Long, *et al.*, "High Speed GaAs Integrated Circuits," Proc. IEEE, **70**(1), pp. 35–45 (1982).
2. J. M. Golio and R. J. Trew, "Optimum Semiconductors for High-Frequency and Low-Noise MESFET Applications," IEEE Trans. Elect. Dev., **30**(10), pp. 1411–1413 (1983).
3. K. Yokoyama, M. Tominzawa and A. Yoshi, "Scaled Performance for Submicron GaAs MESFETs," IEEE Elect. Dev. Lett., **6**(10), pp. 536–538 (1985).
4. M. F. Abusaid and J. H. Hauser, "Calculations of High-Speed Performance for Submicrometer Ion-Implanted GaAs MESFET Devices," IEEE Trans. Elect. Dev., **33**(7), pp. 913–918 (1986).
5. J. M. Golio, "Ultimate Scaling Limits for High-Frequency GaAs MESFETs," IEEE Trans. Elect. Dev., **35**(7), pp. 839–848 (1988).
6. C. E. Weitzel, "Comparison of SiC, GaAs and Si RF MESFET Power Densities," IEEE Elect. Dev. Lett., **16**(10), pp. 451–453 (1995).
7. P. M. Solomon and H. Moroç, "Modulation-Doped GaAs/AlGaAs Heterojunction Field-Effect Transistors (MODFETs), Ultrahigh-Speed Devices for Supercomputers," IEEE Trans. Elect. Dev., **3**(18), pp. 1015–1027 (1984).

8. L. D. Nguyen, L. E. Larsen and U. K. Mishra, "Ultra-High-Speed Modulation-Doped Field-Effect Transistors: A Tutorial Review," Proc. of the IEEE, **80**(4), pp. 494–518 (1992).
9. T. J. Drummond, W. T. Masselink and H. Moroç, "Modulation-Doped GaAs/(Al,Ga)As Heterojunction Field-Effect Transistors: MODFETs," Proc. IEEE, **74**(6), pp. 773–822 (1986).
10. M. Foisy, et al., "The Role of Inefficient Charge Modulation in Limiting the Current-Gain Cutoff Frequency of the MODFET," IEEE Trans. Elect. Dev., **35**(7), pp. 871–878 (1988).
11. S. Tiwari, "Threshold and Sheet Concentration Sensitivity of HEMTs," IEEE Trans. Elect. Dev., **31**(5), pp. 879–887 (1984).
12. S. Bollaert, et al., "Metamorphic InAlAs/OnGaAS HEMTs on GaAs Substrate," IEEE Elect. Dev. Lett., **20**(3), pp. 125–127 (1999).
13. U. Mishra, P. Parikh and Y-F. Wu, "AlGaN/GaN HEMTs – An Overview of Device Operation and Applications," Proc. of the IEEE, **90**(6), pp. 1022–1031 (2002).
14. N. Q. Zang, et al., "High Breakdown GaN HEMT with Overlapping Gate Structure," IEEE Elect. Dev. Lett., **21**(9), pp. 421–423 (2000).
15. Y. Okamota, et al., "High-Power Recessed Gae AlGaN HFET With a Field-Modulating Plate," IEEE Trans. Elect. Dev., **51**(12), pp. 2217–2222 (2004).
16. A. K. Agarwal, "1.1 kV 4H-SiC power MOSFETs," IEEE Elect. Dev. Lett., **18**(7), pp. 586–588 (1997).
17. S.-H. Ryu, et al., "10-kV, 123 mΩ-cm^2 4H-SiC Power DMOSFETs," IEEE Elect. Dev. Lett., 25(8), pp. 556–558 (2004).
18. R. S. Muller and T. I. Kamins, Device Electronics for Integrated Circuits, 2nd edition, John Wiley: New York (1986).
19. T. H. Ning and R. D. Isaac, "Effect of Emitter Contact on Current Gain of Bipolar Devices," IEEE Trans. Elect. Dev., **27**(11), pp. 2051–2055 (1980).
20. G. L Patton, J. C. Bravman and J. D. Plummer, "Physics, Technology, and Modeling of Polysilicon Emitter Contacts for VLSI Bipolar Transistors," IEEE Trans. Elect. Dev., **33**(11), pp. 1754–1768 (1986).
21. T. H. Ning, "History and Future Perspective of the Modern Silicon Bipolar Transistor," IEEE Trans. Elect. Dev., **48**(11), pp. 2485–2491 (2001).
22. T. Izawa, T. Ishibashi and T. Sugeta, "AlGaAs/GaAs Heterojunction Bipolar Transistor," 1985 IEDM Technical Digest, pp. 328–331.
23. F. Zhao, et al., "Analysis of Transit Times and Minority Carrier Mobility in n-p-n 4H-SiC Bipolar Junction Transistors," IEEE Trans. Elect. Dev., **52**(12), pp. 2541–2545 (2005).

24. A. K. Agarwal, *et al.*, "Evolution of the 1600 V, 20 A, SiC Bipolar Junction Transistors," Proceedings of the 2005 IEEE Int. Symp. on Power Semiconductor Devices and ICs, pp. 271–274.
25. Y. Luo, *et al.*, "Fabrication and Characterization of High Current Gain and High Power 4H-SiC Hybrid Darlington Power Transistor," IEEE Trans. Elect. Dev., **51**(12), pp. 2211–2215 (2004).
26. H. Fukui, "Optimal Noise Figure of Microwave GaAs MESFETs," IEEE Trans. Elect. Dev., **26**(7), pp. 1032–1037 (1979).
27. A. Cappy, "Noise Modeling and Measurement Techniques," IEEE Trans. on Microwave Theory and Techniques, **36**(1), pp. 1–10 (1988).

Chapter 4

Optoelectronics

Although there are many different types of optoelectronic devices, we will limit our discussion to three basic technologies: light-emitting diodes, laser diodes, and semiconductor detectors. Despite the fact that this book primarily addresses compound semiconductors, a brief treatment of silicon detectors will also be included because they are widely used for applications with wavelengths below the silicon bandgap edge (1.12 µm), in combination with LEDs or laser diodes made with other materials.

The reliability and radiation response of silicon detectors can be important for some applications of compound semiconductors, including optocouplers and fiber optic systems, and must be considered for overall assessments of reliability and radiation effects on optoelectronics.

4.1 Critical Semiconductor Properties for Light Emission

4.1.1 Recombination and band structure

As discussed in Chapter 2, recombination of electrons and holes can take place through several different processes, including defects within the forbidden region of the bandgap, surface states, recombination in the space charge region of a p-n junction, and direct band-to-band recombination.

For conventional semiconductor diodes the usual goal is to minimize recombination because it is a loss mechanism that interferes with carrier transport. However, for light emitting devices the goal is to make a device with a high recombination rate, where most of the carriers recombine through band-to-band recombination, emitting photons. Another interpretation is that light emitters are lossy devices, where the loss mechanism is intended to maximize light emission.

Band-to-band recombination results in emission of a photon with wavelength

$$\lambda = \frac{hc}{E_g} \qquad (4\text{-}1)$$

where h is Planck's constant, c is the velocity of light, and E_g is the bandgap energy. Equation 4-1 can also be expressed in "practical" units as

$$E = \frac{1.24}{\lambda} \qquad (4\text{-}2)$$

with E in electron volts and λ in μm.

Although band-to-band recombination takes place in all semiconductors, the band-to-band recombination rate for *indirect* semiconductors is 8–10 orders of magnitude lower than for semiconductors with direct bandgaps. Consequently, efficient light emission through band-to-band recombination requires a material with a direct bandgap.

Equation 4-1 implies that there is a single wavelength for band-to-band transitions. However, semiconductors have many possible states, distributed with slightly different energies along the conduction and valence band boundaries. Transitions are possible between the different states, leading to a distribution of photon energies for optical transitions [1], which in turn causes a spread in wavelengths from the photons emitted during band-to-band transitions.

The bandgap energy also depends on doping concentration. For doping levels $>10^{17}$ cm^{-3} the presence of a large number of easily ionized dopant atoms distorts the band structure, causing states with high occupation to spill into part of the normally forbidden region that exists when the semiconductor is lightly doped. This effect is called band tailing, as illustrated in Fig. 4-1. Band tailing is important for laser diodes as well as for light emitting diodes with compensated doping.

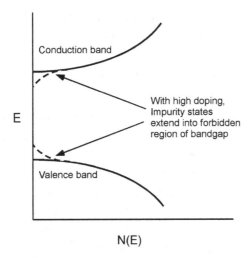

Fig. 4-1. Distortion of band edges with high doping. The extension of impurity bands into the forbidden region allows transitions between lightly populated states in the conduction band and heavily populated states in the valence band, reducing photon energy.

There are several important consequences. First, the photon energy is reduced somewhat compared to lightly doped material because electrons in the conduction band can make a transition to the highly populated "tail" region in the valence band, which extends into the forbidden region of the bandgap.

Second, the wavelength is slightly longer (corresponding to lower energy transitions), reducing absorption losses in the material, and increasing overall efficiency. That characteristic is particularly important for compensated semiconductors.

4.1.2 Recombination processes

For direct bandgap semiconductors, several recombination processes involve absorption or emission of photons:

- Spontaneous emission of a photon, where an electron moves from the conduction band to the valence band,
- Absorption of a photon, elevating an electron to the conduction band and simultaneously creating a hole in the valence band and

- Stimulated emission, where the presence of photons within the semiconductor "triggers" emission of additional photons. *An important property of stimulated emission is that photons produced by this process have the same direction and frequency as that of the initiating photon.*

These three processes are illustrated in Fig. 4-2. The first process, spontaneous emission, is responsible for light emission in light-emitting diodes. The second process (absorption) is the inverse of the spontaneous emission process. The last process, stimulated emission, is the gain mechanism involved in semiconductor lasers because several photons are created by an initiating photon. The probability of stimulated emission is very low under low injection condition, but it becomes the dominant recombination process when the density of photons is sufficiently high. (In a laser, cavity properties also play a role in determining threshold conditions). The fact that the photons produced by that process have the same direction and frequency as the initiating photon is critically important for semiconductor laser operation.

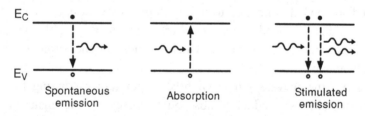

Fig. 4-2. Absorption and emission processes that are accompanied by photon emission or absorption.

Carriers can also recombine through non-radiative processes that do not involve photons. Two important mechanisms are illustrated in Fig. 4-3. Non-radiative recombination can occur through deep-level traps, which are caused by the presence of impurities or crystalline imperfections. This process is shown at the left. In this example the trap can either be neutral, or negatively charged after an electron is captured. We will see later that one of the effects of radiation damage is to increase non-radiative recombination losses through deep-level recombination centers, which are introduced by displacement damage. Degradation from high current or thermal stressing can also cause an increase in non-radiative recombination through bulk and surface traps.

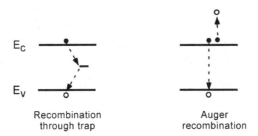

Fig. 4-3. Two important non-radiative recombination mechanisms for direct-bandgap semiconductors.

Energy loss from recombination through traps is usually non-radiative. However, in some semiconductors (notably GaAsP) the addition of certain types of impurities near the band edge can enhance radiative recombination by forming a bound exciton state that first captures an electron, followed by capture of a hole from the valence band. The electron-hole pair then annihilates, emitting a photon with an energy that is slightly lower than the bandgap energy. Absorption losses within the material are reduced for photons with the lower wavelength, which improves the efficiency of light emitters where that process is involved.

Another important loss mechanism is Auger recombination [2,3], shown at the right of Fig. 4-3. In this example, two electrons in the conduction band interact, releasing energy to one of the electrons involved in the initial interaction. The electron with the higher energy becomes a "hot" electron because its energy is so much higher than the mean energy of other electrons in the conduction band. The excess energy is eventually released in the form of heat. The other electron involved in the Auger process loses energy in the initial interaction, falling to the valence band *without releasing a photon* (the lost energy was gained by the "hot" electron in the original interaction). Auger recombination can be considered the inverse of the impact ionization process. High carrier densities are required, as the probability for Auger recombination depends on the third power of the carrier density, as well as on basic material properties. It is particularly important for InP-based semiconductors.

A third loss mechanism, surface recombination, must also be considered. It is an important loss mechanism in III-V semiconductors because it is not possible to form insulators with the same high quality in those materials as silicon dioxide in the silicon system. One way to deal with the poor surface quality is to add a buffer layer to decouple surface recombination from the active device region.

4.1.3 Quantum efficiency

The term quantum efficiency is often used to describe the effectiveness of radiative processes. Internal quantum efficiency, which ignores loss mechanisms in light extraction, is related to the lifetime of non-radiative and radiative processes by the equation

$$Q.E. = \left(1 + \frac{\tau_r}{\tau_{nr}}\right)^{-1} \tag{4-3}$$

where Q.E. is the internal quantum efficiency, τ_r is the radiative recombination lifetime, and τ_{nr} is the non-radiative lifetime. Most optical emitters are designed with internal quantum efficiency >50%.

External quantum efficiency – which essentially depends on the efficiency of extracting photons from the edge or surface – is limited by the high index of refraction of compound semiconductor materials (typically between 3 and 4). As discussed further in Section 4.3, the Brewster angle for rays striking a semiconductor/air interface with a refractive index of ~3.5 is 17° or more. This limits the overall quantum efficiency, particularly for devices that are designed for narrow emission angles, such as LEDs intended for optical fiber applications. Lasers usually have higher external Q.E. than LEDs because most of the photons from stimulated emission travel along the reflective axis. Antireflective coatings are nearly always used to increase light emission efficiency. Lenses, integrated into packaging, can also increase the external quantum efficiency.

4.2 Material Considerations

Several material properties are critical for optoelectronic devices. They include the bandgap, which determines the wavelength as well as electrical properties relating to carrier injection; the index of refraction, that affects photon confinement; the lattice constant and lattice mismatch with other material used in fabrication.

Recombination properties are also important, including the ability to grow material with low defect density, and the Auger coefficient.

4.2.1 Solid solutions

As discussed in Chapter 2, a number of III-V elements can be combined in various proportions to form solid solutions, either as ternary or quaternary material systems. This is called bandgap engineering because it allows the bandgap to be adjusted by suitable selection of materials and proportions of the constituent atoms. For optical applications bandgap engineering provides a way to tailor devices for a range of wavelengths instead of limiting devices for a specific material to a narrow wavelength range, which would be the case if we were forced to limit devices to a single compositional structure and bandgap. The important requirements for the use of III-V solid solutions in heterostructures for optoelecrtronics are as follows:

- The lattice spacing must be closely matched to that of the underlying host material in order to maintain orderly crystal structure and avoid introducing defects. Typically this requires a lattice spacing that is matched within approximately 0.2% (however, strained lattices are an important exception).
- The material must also have a direct bandgap if it is directly involved in light emission or absorption.
- The index of refraction is an important property in optoelectronic devices because of the need to confine photons as well as charge carriers in the overall device structure. Materials with indirect bandgap can be used for photon confinement, and are often incorporated as part of the complex layers that are used in modern optoelectronic devices.

For solid solutions it is important to distinguish between the substrate material and the material combinations used for other layers. This can be very confusing, particularly for quaternary materials. We will use parentheses for the primary material and italics for the substrate in the discussions that follow.

4.2.1.1 Ternary systems using AlGaAs (*GaAs*)

The most widely studied material system for optoelectronics is aluminum gallium arsenide, used for the pioneering work in developing LEDs and laser diodes in the 1970s [4]. Unlike many other material combinations, AlGaAs remains closely lattice matched to GaAs over its entire range of composition, a major advantage. Figure 4-4 shows how the bandgap of solid solutions of AlGaAs changes as the mole fraction, X, of aluminum arsenide is increased. For X = 0.45, the material changes from direct to indirect bandgap, limiting its usefulness for optical transitions to mole fractions below 0.45.

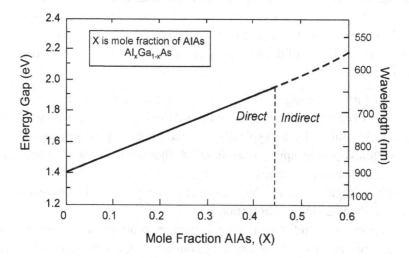

Fig. 4-4. Variation of the energy gap of solid solutions of AlGaAs as the mole fraction of aluminum is increased from 0 (pure GaAs) to 0.45, where the material no longer has a direct bandgap. This corresponds to a wavelength range of 630 to 870 nm.

The bandgap can be adjusted over a range that provides wavelengths from approximately 630 to 870 nm. However, for wavelengths below 700 nm the small separation between the direct and indirect conduction-band minima decreases radiative recombination efficiency [5]. Compressive strain also increases at shorter wavelengths, which reduces the efficiency of AlGaAs at wavelengths <700 nm. Both factors make it more difficult to design devices with wavelengths near the transition region.

4.2.1.2 Quaternary systems using InGaAsP (*InP*)

The fortuitous close lattice matching of AlGaAs over the entire compositional range is one of the reasons that initial work on heterostructures concentrated on that material system. Figure 4-5 shows lattice matching conditions for AlGaAs, as well as more complex quaternary systems using various combinations of InAs, InP, and GaAs. Note the very poor lattice matching for InAs (*GaAs*), the lower curve in the figure.

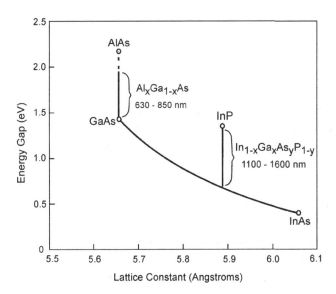

Fig. 4-5. Lattice constant for various material combinations used in optoelectronic devices.

A quaternary system based on InGaAsP (*GaAs or InP*) provides lattice matching over an extended range of energy gaps from about 0.9 to 1.45 eV, as shown by the second vertical line in the figure. This material system has been used to develop LEDs and lasers with wavelengths between approximately 1100 and 1600 nm. Lattice matching considerations restrict the fractional makeup of the constituents. For InGaAsP, the mole fractions of In and Ga are usually interrelated (by [1-X] and X, respectively), with a similar relation between the mole fractions of P and As ([1-Y)] and Y) because those combinations maintain close lattice matching under that condition.

4.2.1.3 Other materials

Several other material systems can be used, including AlGaInP (*InP*) – for wavelengths between 600 and 700 nm – and GaN (on various substrates) that are intended for the 350–480 nm range. We will not discuss their material properties in detail, but refer the reader to the literature for details [6,7].

4.2.2 Strained lattices

Although we have stressed the importance of close lattice matching for heterojunctions, it is possible to grow regions with lattice mismatch up to 2% with low defect density, provided the thickness of the transition region between the two materials is less than about 100 angstroms (see Fig. 2-12, and the related discussion in Chapter 7). The atoms in the transition region are under strain because they are not in their normal position, but are forced into alignment with the underlying material. One advantage of strained lattice technology is that it allows new material combinations to be used, such as InGaAs for lasers and LEDs with wavelengths between 900 and 1100 nm, where there are no suitable materials with close lattice matching.

Another advantage of strained layer technology is that it changes mobility, particularly hole mobility which is very low for most compound semiconductors. This feature can be used to improve the performance of optical emitters. If the strained lattice is designed properly, it can reduce the threshold current density of laser diodes [8]. Many advanced lasers incorporate strained lattices. Despite the lattice mismatch, it is possible to fabricated reliable devices with strained layers as long as the thickness of the strained region is sufficiently low [9].

We will see later that one of the important parameters for semiconductor lasers is the material gain, which effectively determines the increased number of photons that are produced by stimulated emission. Figure 4-6 shows the material gain for strained and unstrained InGaAs [10]. The gain is much higher for the strained material, reducing the minimum current required to initiate lasing. More importantly, the slope is more than five times higher for the strained material, decreasing the current density that is required for a specific lasing condition.

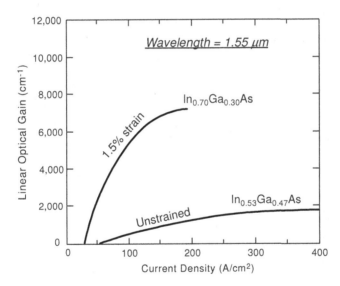

Fig. 4-6. Gain vs. current density for strained and unstrained InGaAs at 1.55 μm [10]. © 1993 IEEE. Reprinted with permission.

4.3 Light-Emitting Diodes

4.3.1 Basic considerations

Light-emitting diodes rely on band-to-band recombination to generate light, with an intensity that is approximately proportional to the forward current through the p-n junction. High efficiency generally requires a material with very low doping in the active region in order to reduce carrier scattering (amphoterically doped structures are an exception). LEDs are analogous to p-i-n diodes, using direct bandgap materials with high probability of band-to-band recombination in the i-region [11].

As discussed earlier, LEDs have a broad spectral width because of the distribution of energies at the bandgap edge. At room temperature the spectral width of typical LEDs is about 5–7% of the peak wavelength. The spectral width increases with injection level because the carrier density increases the number of band tail states. Spectral width will be discussed in more detail in Section 4.4.4.

Figure 4-7 shows the light output of an AlGaAs LED, along with the forward current through the LED as a function of forward voltage. At low current, non-radiative recombination processes dominate, and there is negligible light output.

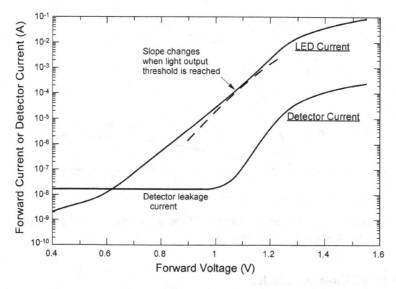

Fig. 4-7. Dependence of optical power and forward current on forward voltage for an LED showing the increase in slope when the forward current is sufficient to overcome non-radiative recombination losses. The detector current is proportional to optical output power.

When the forward voltage reaches about 1.05 V, the injected current is sufficient to overcome non-radiative recombination losses, and there is an increase in the slope of the I–V characteristic (the slope approximately doubles). Light begins to be produced at this transition region. Above that voltage the optical power is proportional to forward current, eventually flattening out because of internal resistance. The region between the change in slope and the saturation point defines the useful range of operating currents for an LED.

Not all LEDs have such a clear distinction between the low injection region where N.R. losses dominate and the region where light emission begins to dominate recombination. If a large number of deep traps are present, the slope can be close to unity even when non-radiative recombination dominates [12].

Although the conversion from current to light is relatively efficient, extraction of light is limited by Snell's law because of the high index of refraction of typical III-V materials, nominally 3.5. As a result, the Brewster angle is about 17° (assuming an air interface), and any photons striking the interface at angles greater than 17° will be reflected back, as shown in the simple diagram of Fig. 4-8. Photons that are produced by spontaneous recombination have arbitrary directions. Consequently, the extracted optical power is about 2% of the electrical power within the LED unless special design techniques are used.

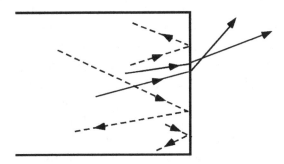

Fig. 4-8. Photons from an LED striking the interface between the III-V material and an interface with n = 1. Spontaneous recombination produces photons with arbitrary direction, most of which strike the interface at oblique angles where they are reflected back into the bulk region and eventually recombine.

Methods to improve light extraction include the use of an integral lens, as well as transparent substrates [13]. Complex designs with internal resonant cavities also can be used to improve light extraction in LEDs with narrow emission angles (*e.g.*, fiber optic applications) [14].

4.3.2 Amphoterically doped LEDs

One of the first processes developed for light-emitting diodes used amphoteric doping [15]. Amphoterically doped LEDs are fabricated with very simple processing techniques, and are still used because of the ease of fabrication and high light-conversion efficiency. The underlying principle for their fabrication is the use of *amphoteric* dopants that can act either as n- or p-type impurities, depending on the growth temperature. LEDs of this type that were fabricated until about 1980 used zinc as an impurity, but that was later supplanted by silicon, which approximately doubles efficiency in GaAs LEDs.

The process starts with an n-doped GaAs wafer. An epitaxial layer is grown using liquified phase epitaxy (LPE). A "melt"–formed by gallium, gallium arsenide and silicon– provides the additional material during the growth process.

The junction is formed by changing the temperature as the epitaxial layer continues to build up. The transition temperature, where the silicon dopant changes from n- to p-type, is approximately 850°C. The main advantage of this process is its simplicity. Doping is only required at the initial processing stage, and devices can be fabricated without the sophisticated processing techniques required for heterostructures.

The high doping level modifies the band structure, producing a large number of band-tail states. Band-to-band transitions can occur between those states, increasing the wavelength compared to lightly doped material. The presence of the band-tail states also reduces recombination near the band edge, increasing the overall efficiency.

It is also possible to grow AlGaAs LEDs with this process by adding aluminum to the melt. The aluminum concentration in this type of AlGaAs LED is lower near the edge of the melt, causing the bandgap to be higher in regions further from the transition region. The gradual change in bandgap within the active part of the junction also reduces recombination, increasing optical conversion efficiency for AlGaAs compared to GaAs LEDs.

Figure 4-9 shows the structure of an amphoterically doped LED. There is a broad transition region between the n- and p-regions of the junction because temperature changes can only occur relatively slowly during the growth process. The junction width is ~20–50 μm [16]. For high efficiency the carrier lifetime must be long enough so that most of the carriers can traverse the junction without undergoing non-radiative recombination, but recombine through radiative recombination. Amphoterically doped LEDs are very sensitive to changes in minority carrier lifetime, which is unusual for most compound semiconductors. This is a direct consequence of the wide junction that is produced by the growth process.

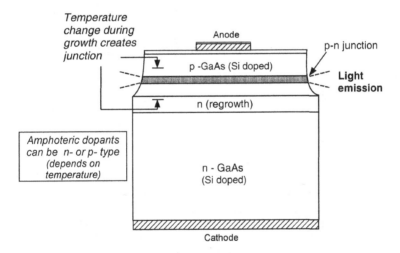

Fig. 4-9. Structure of an amphoterically doped GaAs LED. Although light emission is only shown at the edge, about 30% of the light is emitted from the top surface for typical LEDs of this type.

4.3.3 Heterojunction light-emitting diodes

Heterojunction LEDs use shallow structures, relying on heterostructures that are designed with the correct band offset to allow efficient majority carrier injection over short distances, while still providing sufficient barrier height to avoid back injection of minority carriers. Two basic geometries are used; surface and edge emitters. Surface emitters constrain light operation to a central region, using an etched well through the GaAs substrate that prevents absorption in that region and allows light transmission in the vertical direction. Edge emitters use lateral emission, with a more straightforward design. Both types can be used for optical fiber applications, but edge emitters provide higher bandwidth and more efficient coupling to small-diameter fibers [11].

A diagram of an edge-emitting heterojunction LED is shown in Fig. 4-10. Guiding layers, which have a lower refractive index than the undoped GaAs active layer, are formed next to the active region. They confine most of the photons to the active region. The thickness of the active layer is typically less than 1 μm. LEDs intended for high bandwidth have an even smaller active layer thickness, less than 0.1 μm.

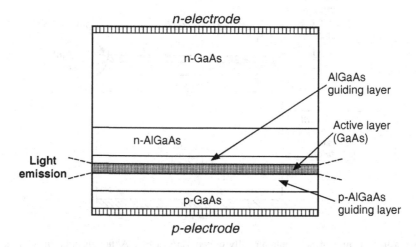

Fig. 4-10. Structure of a double-heterojunction edge-emitting GaAs LED.

The thin active layer causes the angular distribution to be highly asymmetric, with a far-field pattern of ~30° in the plane perpendicular to the junction, and a much narrower pattern in the parallel plane. The narrow light cone emission pattern improves coupling efficiency to optical fibers.

The optical power of this type of LED depends linearly on forward current over several decades. Modulation frequency depends on the thickness of the active layer as well as the injection level. The lifetime is essentially controlled by injected carrier density [17], increasing the frequency response at high current, but also decreasing the carrier lifeitme.

There is a direct tradeoff between those parameters. High-speed LEDs have higher current and photon densities compared to LEDs with lower frequency response.

More complex structures are used for other wavelengths, or for quaternary materials. Metal-organic chemical vapor deposition or liquid-phase epitaxy processing are used to form the thin layers in the LED. The thickness and composition of the layers can be controlled very precisely with those methods. Design details depend on the application. Some LEDs are intended for applications with wide emission angles, making it more difficult to extract light. Several techniques have been developed to optimize light extraction when wide emission angles are needed, including the use of transparent substrates [13] and the use of distributed Bragg reflectors [18].

4.4 Laser Diodes

4.4.1 Basic laser properties

Laser diodes rely on the principle of stimulated emission to provide far higher light generating efficiency than is possible with LEDs. Gain is established by the overall properties of the cavity and the probability of stimulated emission for the semiconductor, which increases with cavity length [19]. The current–power characteristics of a laser are shown in Fig. 4-11.

At low currents the device essentially functions as an LED, with a spectral width that is typically between 5% and 10% of the peak wavelength. The output power is approximately proportional to the input current in this region of operation.

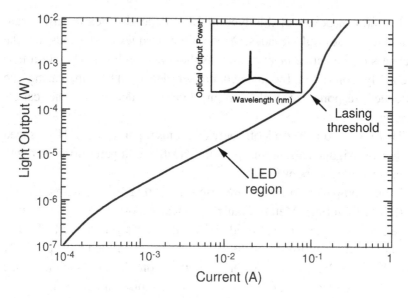

Fig. 4-11. Current-power characteristics of a semiconductor laser showing the transition between the LED and laser operating regions.

Once the lasing threshold is reached, there is a large increase in output power (note the change in slope) as well as a change in the spectral characteristics that is superimposed on the broader spectral width from the LED mode. Further increases in current cause a very large increase in the amplitude of the narrow spectral region, as indicated by the inset in the figure.

Important laser properties include the threshold current where lasing begins; the slope efficiency, which is the differential slope of the optical power output above the threshold current; and the spectral width.

In an edge-emitting laser two parallel surfaces form a Fabry–Perot resonant optical cavity. One facet has a reflective surface, while the other is made partially reflective.

Photons reflected from either facet must travel in a nearly parallel path in order for additional photons generated by stimulated emission to have the preferred direction of the laser cavity. In order for the device to function as a laser, carrier injection must be high enough to provide an overall optical gain that is greater than one. Figure 4-12 shows a simplified diagram of a laser diode. The two guiding regions help to confine photons with more oblique angles to the active region.

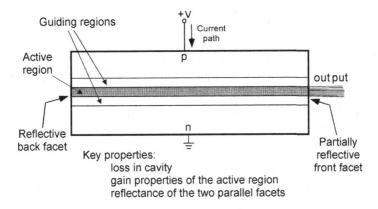

Fig. 4-12. Basic diagram of an edge-emitting semiconductor laser.

Key properties include the reflectivity from the two parallel facets, non-radiative loss within the cavity (including absorption of some photons), and the gain of the semiconductor material. Absorption within the cavity depends on the cavity length, which partially offsets the benefit of a longer cavity for increasing the probability of stimulated emission.

The criteria for lasing can be expressed by

$$\Re \, e^{[(G_{th}-\alpha)L]} = 1 \qquad (4\text{-}4)$$

where $\Re$ is the reflectivity at the partially reflecting surface, G_{th} is the threshold gain per unit length (which depends on the carrier density and material type), α is the absorption coefficient (including various recombination terms, which are lumped into a single term for simplicity), and L is the length of the cavity.

Following the discussion in 4.1.2, three terms are involved in recombination: non-radiative losses, spontaneous emission, and stimulated emission. The latter process produces the additional photons needed for gain. For a semiconductor laser the threshold current corresponds to the condition where the stimulated emission term is just large enough to overcome the loss terms. With internal quantum efficiency χ, the threshold condition can be expressed as

$$\frac{\chi I_{th}}{qV} = (R_{non-rad} + R_{spont} + R_{stim}) \qquad (4\text{-}5)$$

where I_{th} is the threshold current, V is the volume of the laser cavity, $R_{non-rad}$ is the non-radiative recombination term, R_{spont} is the spontaneous recombination term, and R_{stim} is the term corresponding to stimulated radiation. Unless the material gain is very low, the carrier density is essentially clamped at the level corresponding to the threshold even when the current is increased to higher levels. This is the so-called transparency condition.

The recombination terms depend on carrier density, N. For the first two terms on the right of Eq. 4-5 this can be expressed as

$$R_{non-rad} + R_{spont} = (AN + CN^3) + BN^2 \qquad (4\text{-}6)$$

Two terms contribute to non-radiative recombination: bulk recombination centers, proportional to N, and Auger recombination, proportional to N^3. The spontaneous recombination term is proportional to N^2 with bimolecular recombination coefficient, B. For the materials that we are considering, the magnitude of B is $\sim 10^{-10}/cm^3$.

It is important to design lasers that operate with a carrier density that is low enough to keep the Auger loss term low (it depends on N^3), particularly for InGaAsP which has a higher Auger coefficient than AlGaAs.

The carrier density corresponding to the laser threshold depends on the gain of the semiconductor material system, which is also affected by the design of the laser cavity. Figure 4-6 (Section 4.2.2) showed how the material gain can be improved by using strained lattices. Quantum-well confinement – which requires several layers with precisely controlled thickness within the active region of the laser – provides another way to reduce loss and improve material gain, at the expensive of more complicated processing. Strained layers and quantum confinement are frequently used in commercial lasers, particular those using quaternary compounds with nominal wavelengths of 1300 and 1500 nm.

Figure 4-13 shows the approximate dependence of material gain on current for GaAs and InGaAs heterostructures.

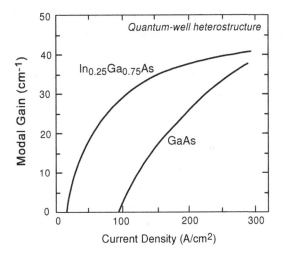

Fig. 4-13. Dependence of modal gain on current density for quantum-well lasers fabricated with InGaAs and GaAs [8]. © 2000 IEEE. Reprinted with permission.

The sharp increase in modal gain occurs at a much lower current density for InGaAs, with a steeper slope. This allows lasers to be fabricated with lower threshold current in that material.

The typical carrier density where the gain begins to increase is on the order of 10^{18} cm^{-3} for most of the materials that we are considering. Laser design equations take the shape of the modal gain curve into account. A logarithmic dependence of gain on carrier density has been shown to be applicable to several material systems using conventional as well as quantum-well lasers [20].

Laser properties, including the threshold current, are temperature dependent. A simplified way of evaluating this uses the coefficients of the various recombination terms in Eq. 4-6, resulting in the following expression for the threshold carrier density, N_{th} [21]:

$$N_{th} = N_i \left(1 + \frac{A + CN_i^2}{BN_i} \right) \qquad (4\text{-}7)$$

with N_i the intrinsic carrier density. The first-order dependence on temperature is due to the bimolecular recombination coefficient, B, which varies as $T^{-3/2}$. That coefficient determines the number of photons that recombine through spontaneous recombination, which must be present in sufficient numbers to initiate stimulated recombination before the lasing condition can be reached.

The other factor is the Auger recombination term, C, which is about one order of magnitude higher for InGaAsP compared to GaAs. The Auger term can cause saturation of the power output when InGaAsP lasers are operated at high output power.

Typically the threshold current increases by about a factor of three at 100°C compared to its value at 20°C. Slope efficiency also decreases, but the effect of temperature on slope efficiency depends on the design of the cavity (it is lower for strained quantum-well lasers [22]).

The temperature dependence is more complicated for InGaAsP because of the larger Auger coefficient. The carrier density will increase with temperature because of the decrease in the B term, but if the carrier density increases sufficiently the Auger term – which increases *quadratically* with carrier density – will also increase rapidly, causing the temperature dependence to increase in a highly nonlinear way [23]. For this reason, AlGaAsP lasers are designed to operate at low carrier densities to avoid the overwhelming influence of the Auger term at high carrier densities.

Temperature dependence of the threshold current is typically characterized by the characteristic temperature, T_o, defined by

$$T_o = J_{th} (dJ_{th}/dT)^{-1} \tag{4-8}$$

Devices with high T_o have lower temperature sensitivity. Typical values for GaAs and AlGaAs lasers are 150–200. T_o values for InGaAsp and InGaAs are much lower, between 50 and 70. Recent work using pressure to increase the band gap has shown that the increased temperature sensitivity of long-wavelength lasers is dominated by non-radiative processes, such as Auger recombination, even for 1.3-µm lasers with very low threshold current [24].

4.4.2 In-plane semiconductor lasers

Modern laser diodes may have numerous layers. They are used to increase efficiency, decrease surface recombination, and improve light confinement. A diagram of a basic five-layer in-plane laser is shown in Fig. 4-14. Current flow takes place between the contact at the top and the bottom contact to the n-GaAs substrate.

In this example carrier flow is restricted by lateral oxide regions at the surface, as shown. The lightly doped active region is sandwiched between two guiding regions that have a lower refractive index than the GaAs active layer. Other methods can be used to isolate the flow of carriers and to define the active region of the laser, including v-grooves formed by selective etching, proton bombardment to modify the properties of the top layers, and using additional layers to produce buried heterostructures.

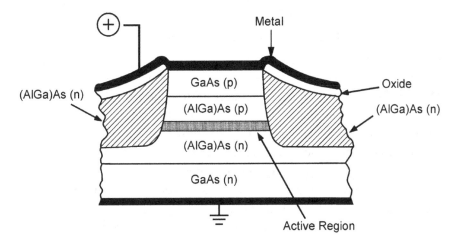

Fig. 4-14. Diagram of a basic five-layer laser.

As discussed earlier, more sophisticated structures are commonly used, including multiple quantum-well lasers where several very thin layers are present in the active region that allow quantum-layer confinement. Quantum wells can reduce the threshold current by a factor of five or more. Figure 4-15 shows the layers in a 1550-nm laser with four different quantum levels. The mole fraction of Al is different for successive layers. The thickness and spacing of each layer must be precisely controlled. The resulting device has a lower threshold current and higher efficiency, a major advantage for InGaAsP lasers.

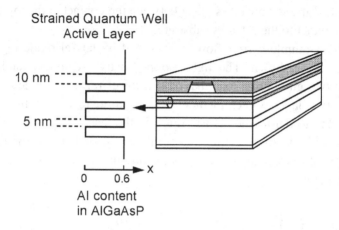

Fig. 4-15. Layer structure in a four-level quantum-well laser diode.

Figure 4-16 shows the dependence of power output on current for a 1550 nm laser diode using a linear scale. Although not evident from this figure, there is significant light output well below the threshold current, where the laser operates as an LED. However, once the threshold condition is reached there is a sudden change in output power, as shown by the inset in the figure. As noted earlier, the spectral width is very narrow when the device reaches the lasing condition compared to the wider spectral width in the LED mode.

The threshold current increases with increasing temperature, affecting the stability of the laser output power. This can be accommodated within narrow limits, which can be done by mounting the laser on a small thermoelectric cooler with a feedback circuit to control the temperature. Alternatively, the photocurrent from an internal monitor diode can be used as part of a feedback circuit that maintains constant optical power output if the temperature changes. The slope efficiency – the slope above the threshold current – remains relatively constant except at high power levels, where internal resistance and self-heating cause the slope to decrease.

Optoelectronics 99

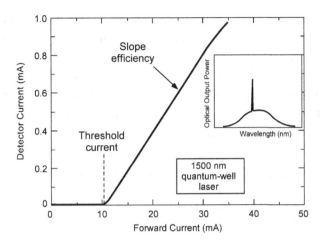

Fig. 4-16. Dependence of optical power on forward current for a 1550-nm quantum-well laser diode.

The derivative of the power output with respect to forward current is often used as a figure of merit, or as a diagnostic parameter, for laser diodes. Figure 4-17 shows a derivative plot for a laser where the slope efficiency decreases with increasing current.

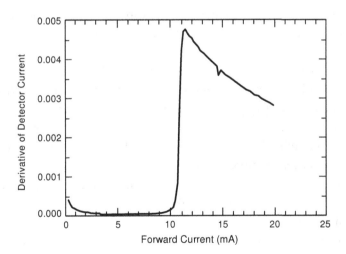

Fig. 4-17. Derivative of optical power for a laser diode where the slope efficiency decreases with current. This is a useful diagnostic method for laser diode performance.

Using the derivative is a more effective way to determine how reliability-related stress or radiation damage affects laser diodes compared to normal I–V curves [25]. However, measurements of the optical power must be made with small incremental current steps and high precision to avoid steps and discontinuities when the derivative is calculated. The "glitch" in Fig. 4-17 at a forward current of 14.6 mA is due to measurement inaccuracy.

4.4.3 Vertical cavity semiconductor lasers (VCSELs)

Vertical cavity semiconductor lasers use Bragg reflectors to form a vertical Fabry–Perot resonant cavity [26]. The top surface is partially reflecting, allowing some of the light to be emitted vertically. The reflectors are formed by as many as 70 different layers, spaced at distances of a quarter of the wavelength. The reflective properties are obtained by using thin structures with different doping levels, which changes the index of refraction. Current flows through the Bragg reflectors as well as through the body of the structure. For this design, one Bragg assembly is p-type, while the other is n-type, requiring different dopant atoms in the two reflecting structures. The thickness of the layers in the Bragg reflectors must be precisely controlled. Several modes are usually present in VCSEL structures, partly because of small deviations in layer thickness.

A diagram of one implementation of a VCSEL is shown in Fig. 4-18. The active region is limited to the smaller cylinder shown in the diagram by an oxide layer that restricts current flow to the center, increasing efficiency. The active region is very thin; most of the volume is required to form the complex Bragg reflectors. VCSELs have several advantages, particularly for applications that do not require a large amount of optical power. They provide vertical light emission, making it easier to couple the emitted light to detectors (or other assemblies), as well as very low threshold current.

Thermal effects play an important role in VCSELs. Heat transfer between an in-plane laser and the substrate is highly efficient, allowing that type of laser to operate over a relatively broad range of currents.

Thermal dissipation in a VCSEL is much less efficient because the active part of the device is effectively isolated by the multiple layers of the reflector. Consequently, VCSELs operate over a relatively restricted range of currents because of self-heating [27].

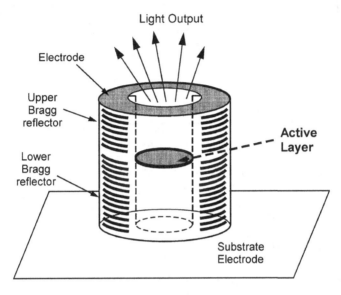

Fig. 4-18. Diagram of a VCSEL structure showing the two Bragg reflectors and the very thin active region at the center.

Figure 4-19 shows the optical power characteristics of a VCSEL vs. forward current. Unlike in-plane lasers, the slope efficiency begins to change at currents that are about twice the threshold current due to the temperature rise in the active region. This restricts the operating region of a VCSEL to a relatively narrow range, typically less than three times the threshold current.

VCSELs can be produced at many different wavelengths. The first VCSELs used AlGaAs (*GaAs*) at 850 nm, but they are now available at wavelengths of 650 to 700 nm as well as for the second fiber-optic window (1300 nm) using different materials.

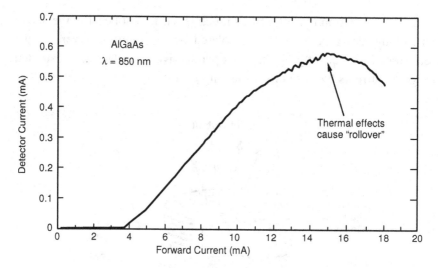

Fig. 4-19. Optical power vs. forward current for an 850-nm VCSEL.

4.4.4 Spectral width

4.4.4.1 Light-emitting diodes

The spectral width of an LED can be estimated by assuming a parabolic E-k energy band distribution near the band edge. With this assumption, the spectral width has a Lorentzian lineshape that is affected by homogeneous broadening as well as the distribution of energy values at the edge of the bandgap. The peak wavelength decreases with temperature due to the temperature dependence of the energy gap. The spectral width increases slightly with temperature as the distribution of energy states is also temperature dependent.

Light *emission* from an LED is more complicated because the emission and absorption spectra have a slightly different dependence on wavelength. Part of the intrinsic spectrum will be absorbed as the light travels through the material before it exits from the window. Figure 4-20 shows how the spectrum of a GaAs LED (λ = 879 nm) is altered when it passes through a 380-µm n-doped GaAs wafer ($N_D = 10^{17}$ cm^{-3}) [28]. The peak wavelength has shifted from 879 to 905 nm due to absorption by the GaAs region. In this study the thickness of the GaAs decreased the overall transmission to 4.5%. The spectral width of this LED is about 5% of the peak wavelength, which is typical of many LEDs.

One way to improve LED performance is to use a material where there is a larger shift between the emission and absorption spectra. This is the reason for the high efficiency of GaAsP LEDs, where the isotonic impurities involved in light emission are located slightly below the bandgap.

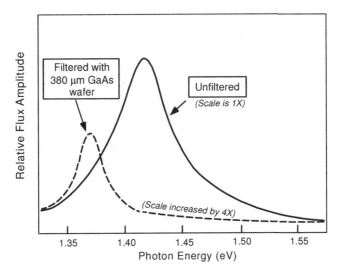

Fig. 4-20. Comparison of unfiltered and filtered spectra from a GaAs LED showing the effect of internal absorption on the spectrum [28]. © 1965 IEEE. Reprinted with permission.

4.4.4.2 In-plane lasers with Fabry–Perot reflectors

The spectral width of a laser is considerably smaller than that of an LED, but depends on the specific design of the laser. The spectral width of a basic F–P laser is determined by the laser cavity properties as well as by the properties of the semiconductor. We will define the laser cavity efficiency, $\gamma(R)$, by

$$\gamma(R) = \frac{1}{R}\left(\frac{1-R}{\ln(1/R)}\right) \qquad (4\text{-}9)$$

where R is the reflectivity of the partially reflecting facet. Absorption losses are not included in this equation. Note that γ is near unity for a well defined laser.

The spectral half with, Δλ, is given by [29]

$$\Delta\lambda = \frac{c\pi(h\nu)}{2P_o} \ln\left(\frac{1}{R}\right)\gamma(R)\left(\frac{\Delta\lambda}{\lambda}\right)^2 K \qquad (4\text{-}10)$$

where hν is the photon energy, P_o is the total power emitted per facet, and K is an astigmatism factor that depends on the details of the laser geometry. The essential point of this equation is that the spectral width depends inversely on the optical power level, and to a lesser degree on the cavity properties.

Experimental results for a V-groove AlGaAs laser (820 nm) are shown in Fig. 4-21 for an optical power level of 5 mW per facet [29]. At higher power levels Δλ is about 10 Å; approximately 0.1% of the center wavelength.

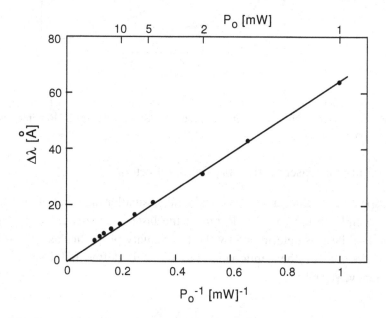

Fig. 4-21. Dependence of spectral width of a F-P laser on optical power [29]. © 1983 IEEE. Reprinted with permission.

4.4.4.3 Laser with distributed feedback reflectors

A distributed feedback reflector alters the cavity characteristics, increasing the cavity gain for specific mode conditions. Such lasers have a spectral wavelength that is theoretically narrower than those with F-P reflectors, but the dependence of the wavelength on optical power is more complicated. VCSELs are an example of a laser with a DBR reflector, although most VCSELs are not designed to operate with a narrow wavelength. It is possible to obtain single-mode operation in lasers with Bragg reflectors.

The linewidth is affected by fluctuations of the phase of the optical fields within the cavity, which increases the linewidth broadening above that expected from homogeneous broadening. The linewidth, Δv, of a single-mode laser is [30]

$$\Delta v = \frac{K}{4\pi S}(1+\alpha^2) \qquad (4\text{-}11)$$

where K is the average spontaneous emission rate (normalized to the cavity volume), S is the photon density within the cavity, and α is the linewidth enhancement factor that incorporates the phase fluctuations.

In order to achieve single-mode operation, the power level must be low enough to suppress other modes within the cavity. This is easier to achieve with a long cavity length, although that increases the threshold current and lowers quantum efficiency. From the standpoint of reliability, the conditions for single-mode operation are affected by cavity loss, and may change if radiation or reliability mechanisms increase non-radiative losses within the relatively long cavities.

A study of the design of a single-mode InGaAsP laser by Takahashi, *et al.* [31] used the product of the linewidth and optical power as a figure of merit. They showed that the linewidth was close to the theoretical limit of 3.2 MHz for a laser cavity length of 200 µm. That corresponds to a fractional wavelength of 1.5×10^{-8} for the 1.5 µm wavelength of the lasers in the study. This illustrates the extreme precision that can be obtained for single-mode lasers.

4.4.5 Tunable lasers

Another important class of lasers is the tunable laser, where the design incorporates methods of changing the frequency. The goal is often to achieve a wide tuning range for multiplexing applications. There are many ways of implementing tuning including [32]:

- Use of selectable distributed feedback arrays;
- Incorporation of an external cavity, where the cavity properties are affected by other mechanisms;
- Use of a movable element that is integrated within the laser, such as a MEMS structure;
- Grating coupled sampled reflectors;
- Sampled grating distributed Bragg reflectors with an integrated optical amplifier.

A recent study used a VCSEL structure with an internal heater that provided a tuning range of 7% by changing the temperature of the VCSEL, which operated at a nominal wavelength of 780 nm [33]. The wavelength change was caused by the effect of temperature on the refractive index. The advantage of this design is its simplicity and compactness.

We will not discuss tunable lasers in detail, but note that reliability and radiation damage depend on the specific design and the method used for tuning. An example of a reliability study of a tunable laser is included in Section 7.5.2.

4.5 Detectors

4.5.1 Light absorption

Light is absorbed exponentially within a semiconductor, as described by Beer's law

$$I = I_o e^{-\alpha x} \qquad (4\text{-}12)$$

where I is the light intensity at a distance x in the material, I_o is the light intensity at the surface, and α is the absorption coefficient in cm^{-1}. The reciprocal of the absorption coefficient is an approximate measure of the thickness required for light absorption. For an indirect material such as silicon, the absorption coefficient has a relatively broad dependence on wavelength, as shown in Fig. 4-22. Consequently it is necessary to use relatively thick layers to absorb light at longer wavelengths.

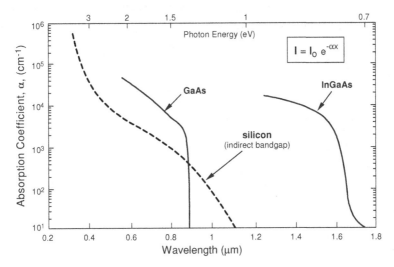

Fig. 4-22. Dependence of the absorption coefficient on wavelength for several semiconductors.

For materials with direct bandgap, the absorption coefficient has a more abrupt dependence on wavelength. The slope is not quite infinite due to the parabolic dependence of quantum states within the crystal, but it is much steeper than for an indirect bandgap material.

The steep dependence of the absorption coefficient on wavelength allows light to be absorbed by relatively thin layers for compound semiconductors compared to silicon. That property, along with the ability to use a complex stack of thin layers and bandgap engineering, makes it possible to use compound semiconductors to design photodetectors with high bandwidth and shallow absorption depth.

4.5.2 Basic p-n junction photodetector

A diagram of a p-n junction photodetector, under reverse bias for operation as a current detector, is shown in Fig. 4-23. Two different detector types are shown, representative of silicon technology. The first uses regions with moderate doped levels. For that structure, photocurrent is collected by drift for photons generated within the depletion region of the junction, as well as by diffusion for photons that are generated beyond the depletion region. The second detector uses an undoped intrinsic region, biasing the device with sufficient voltage to allow the entire i-region to be depleted. In this case nearly all the photocurrent is collected by drift.

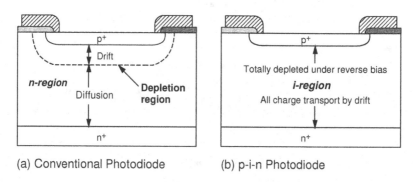

Fig. 4-23. Comparison of conventional and p-i-n silicon photodiodes where the low value of the absorption coefficient requires thick regions for light absorption.

For silicon, the thickness of the region must be compatible with the intended wavelength. Thicker regions are required at long wavelengths because of the low value of the absorption coefficient for wavelengths >850 nm (the light intensity decreases by a factor of 1/e in a distance of $1/\alpha$). For high-speed applications the capacitance and charge collection time are important [34]. The charge collection time is affected by drift through the depletion region as well as diffusion.

An InGaAs p-i-n diode is shown in Fig. 4-24 that can be used for wavelengths up to 1.7 µm. The light absorbing layer is much thinner because of the direct bandgap of InGaAs that allows nearly all of the light to be collected in a shallow region. Note that this detector is inverted compared to the silicon devices in Fig. 4-23, with light entering from the back.

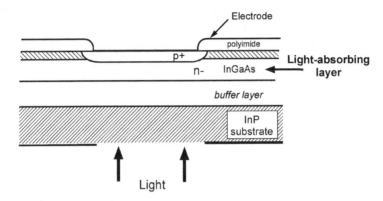

Fig. 4-24. Diagram of an InGaAs (InP) p-i-n detector [35]. The structure is inverted compared to the silicon detectors in Fig. 4-23. © 1995 IEEE. Reprinted with permission.

For wavelengths above 1 μm very little light is absorbed within the InP substrate, allowing it to be interposed between the active part of the device and the entrance region with minimal effect on light transmission. However, the strong absorption in InP at wavelengths <1 μm cuts off absorption at shorter wavelengths compared to an InGaAs photodiode with absorption from the top surface [35].

The most important detector properties are wavelength range, which affects the thickness of the absorbing region as well as the design of antireflective coatings; the responsivity, which depends on wavelength and is usually expressed in A/W; leakage current; and dark current. Dark current is particularly important for p-i-n detectors. It is temperature dependent, and is also affected by the density of bulk defects in the intrinsic region.

We will see in Chapter 12 that one of the main effects of radiation damage is to increase dark current in photodetectors because of bulk defects that are introduced by radiation. Bulk and surface defects are also important for reliability, as discussed in Chapter 7.

4.5.3 Avalanche photodiodes

Avalanche photodiodes (APDs) use a high-field region to increase the number of photons collected from incident light through avalanche multiplication. The basic design consists of a light collection (drift) region, followed by a high-field region where carriers – usually electrons, because of higher mobility – are multiplied by the avalanche multiplication process. A simplified diagram of a silicon APD is shown in Fig. 4-25.

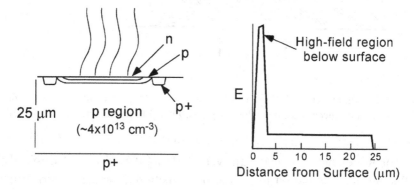

Fig. 4-25. Diagram of a basic silicon avalanche photodiode, along with the electric field profile.

In this example the high-field region is just below the surface. Channel stops around the active region reduce surface leakage current. The right part of the figure shows the electric field in the two regions. The device is biased to provide a specific value of multiplication in the high-field region, typically between 3 and 30. Dark current in the drift region is also affected by the multiplication factor, but surface leakage current is not.

APDs provide better signal-to-noise ratios than conventional detectors, but the bias conditions must be carefully controlled, usually optimizing them for specific devices. The avalanche properties are also affected by temperature, further complicating the way that bias conditions are regulated in applications.

An example of an InGaAs APD is shown in Fig. 4-26 [36]. More regions are present in this structure compared to the APD in Fig. 4-24, but the basic principles are the same. The p+ region below the contact is used to reduce surface leakage. It extends laterally to form a guard region, as shown. Photons are absorbed in the n–InGaAs region, producing carriers that are multiplied in the n–InP layer. A grading layer of InGaAsP is used to increase the bandgap discontinuity between the InGaAs absorption layer and the InP multiplication region to reduce dark current. This illustrates the complexity of high-performance detectors. Fabrication details, including the composition and thickness of the various layers, are often proprietary.

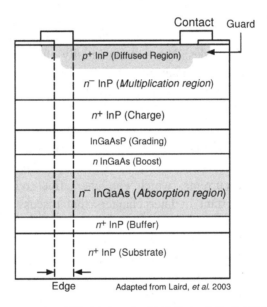

Fig. 4-26. A more complex APD fabricated with InGaAs [36]. © 2003 IEEE. Reprinted with permission.

There are many applications of photodetectors. One of the most important is in high-bandwidth optical receivers. Noise and frequency response are key properties for such applications. Further details are discussed in the literature [37], including tradeoffs between p-i-n and avalanche photodetectors.

4.6 Summary

This chapter has discussed the basic properties of LEDs, laser diodes, and detectors. Considerable emphasis was placed on the interplay between the design, material properties, and critical parameters of the various devices, which will be referred to in later chapters that discuss reliability and radiation effects.

LEDs are often used in high-reliability applications, and it is important to understand the different ways in which LEDs are designed, as well as the reasons that the current density and lifetime are so different for various LED technologies. Amphoterically doped LEDs require very long lifetimes in order to operate efficiently, which is very unusual for compound semiconductor devices. On the other hand, high-speed LEDs have extremely thin active layers, with current densities that are nearly as high as that of laser diodes. This can potentially affect their reliability, which will be discussed in Chapter 6.

Different material properties are important for lasers. The high Auger recombination coefficient of InGaAsP makes it critically important to design lasers in that material that can operate at carrier densities below the point where they will be overwhelmed by the loss terms from Auger recombination. For different reasons, VCSELs can only operate over a restricted range of currents, which can also affect reliability. These factors will be important in later chapters that discuss reliability and radiation damage.

The physical design of a laser is also important. Although many lasers use basic Fabry-Perot cavities, Bragg reflectors are used in VCSELs as well as in special frequency-stabilized lasers.

Finally, detectors are required for nearly all applications of photonics. The brief discussion of detector technologies in this chapter provides the background to understand how reliability and radiation damage affect their properties in later chapters.

References

1. M. B. Panish, "Heterostructure Injection Lasers", Proc. of the IEEE, **64**(10), pp. 1512–1540 (1976).

2. A. Sugimura, "Comparison of Band-to-Band Auger Processes in InGaAsP", IEEE J. Quant. Elect., **19**(6), pp. 930–932 (1983).
3. R. Fehse, *et al.*, "A Quantitative Study of Radiative, Auger, and Defect Related Recombination Processes in 1.3 μm GaInNAs-Based Quantum-Well Lasers", IEEE J. on Selected Topics in Quant. Elect., **8**(4), pp. 801–810 (2002).
4. M. Razeghi, "Optoelectronic Devices Based on III-V Compound Semiconductors Which Have Made a Major Scientific and Technological Impact in the Past 20 Years", IEEE J. on Selected Topics in Quant Elect., **6**(6), pp. 1344–1354 (2000).
5. S. Yamamoto, *et al.*, "680 nm CW Operation at Room Temperature by AlGaAs Double Heterojunction Lasers", IEEE J. Quant. Elect., **19**(6), pp. 1009–1015 (1983).
6. E. P. O'Reilly and A. R. Adams, "Band Structure Engineering in Strained Semiconductor Lasers", IEEE J. Quant. Elect., **30**(2), pp. 366–379 (1994).
7. S. P. Denbaars, "Gallium-Nitride-Based Materials for Blue to Ultraviolet Optoelectronics Devices", Proc. of the IEEE, **85**(11), pp. 1740–1749 (1997).
8. J. J. Coleman, "Strained-Layer InGaAs Quantum-Well Heterostructure Lasers", IEEE J. on Selected Topics in Quant. Elect., **6**(6), pp. 1008–1013 (2000).
9. I. J. Fritz, *et al.*, "Critical Layer Thickness in $In_{0.2}Ga_{0.8}As$/GaAs Single Strained Quantum Well Structures", Appl. Phys. Lett., **51**(13), pp. 1004–1006 (1987).
10. C. H. Lin and Y. H. Lo, "Empirical Formulas for Design and Optimization of 1.55 μm InGaAs/InGaAsP Strained-Quantum-Well Lasers", IEEE Phot. Tech. Lett., **5**(3), pp. 288–290 (1993).
11. D. Marcuse, "LED Fundamentals: Comparison of Front- and Edge-Emitting Diodes", IEEE J. Quant. Elect., **13**(10), pp. 819–827 (1977).
12. A. A. Bergh and P. J. Dean, "Light-Emitting Diodes", Proc. of the IEEE, **60**(2), pp. 156–223 (1972).
13. D. A. Vanderwater, *et al.*, "High Brightness AlGaInP Light Emitting Diodes", Proc. of the IEEE, **85**(11), pp. 1752–1764 (1997).
14. D. Delbeke, *et al.*, "High-Efficiency Semiconductor Resonant-Cavity Light-Emitting Diodes: A Review", IEEE J. Selected Topics on Quant. Elect., **8**(2), pp. 189–206 (2002).
15. H. Kressel and M. Ettenberg, "Electroluminescence and Photoluminescence of GaAs:Ge Prepared by Liquid Phase Epitaxy", Appl. Phys. Lett., **23**(9), pp. 511–513 (1973).

16. H. Rupprecht, et al., "Efficient Electroluminescence from GaAs Diodes at 300°K", Appl. Phys. Lett., **9**(6), pp. 221–223 (1966).
17. M. Ettenberg, K. C. Hudson and H. F. Lockwood, "High Radiance Light Emitting Diodes", IEEE J. Quantum Elect., **12**(6), pp. 360–364 (1976).
18. P. Royo, R. P. Stanley and M. Ilegems, "Analytical Calculations of the Extraction Efficiency of Microcavity Light-Emitting Diodes for Display and Fiber Coupling Applications", IEEE J. on Selected Topics in Quant. Elect., **8**(2), pp. 207–218 (2002).
19. M. G. A. Bernard and G. Duraffourg, "Laser Conditions in Semiconductors", Phys. Status Solidi, **1**, pp. 699–703 (1961).
20. T. A. DeTemple, "On the Semiconductor Laser Logarithmic Gain-Current Density Relation", IEEE J. Quant. Electron., **29**(5), pp. 1246–1252 (1993).
21. A. Haug, "Theory of the Temperature Dependence of the Threshold Current of an InGaAsP Laser", IEEE J. of Quant. Elect., **21**(6), pp. 716–718 (1985).
22. R. J. Fu, et al., "High-Temperature Operation of InGaAs Strained Quantum-Well Lasers", IEEE Phot. Tech. Lett., **3**(4), pp, 308–310 (1991).
23. A. F. Phillips, et al., "The Temperature Dependence of 1.3- and 1.5-μm Compressively Strained InGaAs(P) MQW Semiconductor Laser", IEEE J. of Selected Topics in Quant. Elect., **5**(3), pp. 401–412 (1999).
24. I. P. Marko, et al., "Recombination and Loss Mechanisms in Low-Threshold InAs-GaAs 1.3-μm Quantum-Dot Lasers", IEEE J. on Selected Topics in Quant. Elect., **11**(5), pp. 1041–1047 (2005).
25. I. Vurgaftman and J. R. Meyer, "Effects of Bandgap, Lifetime, and Other Nonuniformities on Diode Laser Thresholds and Slope Efficiencies", IEEE J. on Selected Topics in Quant. Elect., **3**(2), pp. 475–484 (1997).
26. K. Iga, "Surface-Emitting Laser – Its Birth and Generation of a New Optoelectronics Field", IEEE J. on Selected Topices in Quant. Elect., **6**(6), pp. 1201–1215 (2000).
27. K. D. Choquette and H. Q. Hou, "Vertical-Cavity Surface Emitting Lasers: Moving from Research to Manufacturing", Proc. IEEE, **85**, pp. 1730–1747 (1997).
28. W. N. Carr, "Characteristics of a GaAs Spontaneous Infrared Source with 40 Percent Efficiency", IEEE Trans. Elect. Dev., **12**(10), pp. 531–535 (1965).
29. G. Arnold, K. Petermann and E. Schlosser, "Spectral Characteristics of Gain-Guided Lasers", IEEE J. Quant. Elect., **19**(6), pp. 974–980 (1983).
30. C. H. Henry, "Theory of the Linewidth of Semiconductor Lasers", IEEE J. Quant. Elect., **18**(2), pp. 259–264 (1982).

31. M. Takahashi, et al., "Narrow Spectral Linewidth 1.5 μm GaInAsP.InP Distributed Bragg Reflector (DBR) Lasers", IEEE J. of Quant. Elect., **25**(6), pp. 1280–1287 (1989).
32. L. Coldren, et al., "Tunable Semiconductor Lasers: A Tutorial", IEEE. J. Lightwave Technology, **22**(1), pp. 193–202 (2004).
33. S.-S. Yang, et al., "Wavelength Tuning of Vertical cavity Surface Emitting Lasers by an Internal Device Heater", IEEE Phot. Tech. Lett., **20**(20), pp. 1679–1681 (2008).
34. L. K. Anderson and B. J. McMurtry, "High-Speed Photodetectors", Proc. IEEE, **54**(10), pp. 1335–1348 (1966).
35. Y.-G. Wey, et al., "110-GHz GaUbAs/InP Double Heterostructure p-i-n Photodetectors", IEEE J. Lightwave Tech., **13**(7), pp. 190–1499 (1995).
36. J. S. Laird, et al., "Heavy-Ion Induced Single-Event Transients in High-Speed InP-InGaAs Avalanche Photodiodes", IEEE Trans. Nucl. Sci., **50**(6), pp. 2225–2232 (2003).
37. M. C. Brain, "Comparison of Available Detectors for Digital Optical Fiber Systems for the 1.2–1.55 μm Wavelength Range", IEEE J. Quant. Electron., **18**(2), pp. 219–224 (1982).

Chapter 5

Reliability Fundamentals

This chapter discusses basic reliability concepts, providing a necessary background for understanding the reliability of specific devices and their failure mechanisms in later chapters. In most cases reliability is treated as a statistical problem, using a mathematical relationship to approximate the actual failure distribution that takes place when devices are subjected to stress and/or operated for an extended time. Reliability testing is usually done with a relatively small test sample, using the results of the sample test to make a statistical estimate of the overall reliability and failure rate of the parent population.

Devices used in electronic assemblies nearly always have very low failure rates, a basic requirement for applications in any system that is expected to operate over periods of several years. The low failure rate of individual devices makes it necessary to perform reliability tests under stress conditions that accelerate the failure rate in order to complete the tests in reasonable time periods. Identifying the appropriate stress conditions and the assumptions needed to extrapolate test results to the far lower stress conditions that represent actual use conditions are fundamental to any reliability study.

We need to keep in mind that although sample testing and statistical models are widely used, conclusions based on sample testing ultimately depend on how well the actual failure distribution of the parent population obeys the function that is assumed for the overall failure distribution, as well as assumptions about the acceleration factors.

Several examples will be shown in later chapters where the actual effect of accelerated stress conditions is different when it is examined more closely, changing the conclusions about sample tests that are used to verify part reliability.

5.1 Reliability Requirements

5.1.1 Definitions

The reliability of a population of devices is defined as the fractional number of devices in the total population that continue to survive after operating for a time, t, compared to the initial population

$$R(t) = \frac{N_{surv}}{N_{initial}} \qquad (5\text{-}1)$$

where R(t) is the reliability, N_{surv} is the number of surviving devices after an elapsed time interval under operational or stressed conditions, and $N_{initial}$ is the number of devices in the initial population.

The number of failed devices, F(t), is given by

$$F(t) = 1 - R(t) \qquad (5\text{-}2)$$

Another parameter used in reliability is the failure density function, f(t), which is the derivative of F(t). The dimension of f(t) is $(t)^{-1}$; f(t) is not normalized and can have very high values for some failure conditions.

A related parameter is the failure rate, $\lambda(t)$, which is just f(t) normalized by the total reliability, R(t)

$$\lambda(t) = \frac{f(t)}{R(t)} = \frac{f(t)}{1 - F(t)} \qquad (5\text{-}3)$$

F(t), f(t) and $\lambda(t)$ are interrelated, with a dependence that is a function of the failure distribution. Although many possible functions can be used to describe the failure density function f(t), much of the work on reliability indicates that the log-normal function is the most appropriate, particularly for wearout-related mechanisms where f(t) is expected to increase with time. However, other statistical distributions may be more appropriate for cases where infant mortality causes the failure rate to be higher at shorter times.

5.1.2 Properties of probability distributions

One way to select an appropriate distribution function is to compare its basic properties with the properties of the test sample, particularly the failure rate. We will review the properties of three statistical distribution functions that are often used for failure analysis: the exponential, log normal, and Weibull distributions.

5.1.2.1 Exponential distribution

The probability density of the exponential distribution is described by

$$f(t) = \lambda e^{-\lambda t} \qquad (5\text{-}4)$$

For this distribution the failure rate, λ, is constant. It is appropriate for random failures, but not for wearout effects because the constant failure rate is inconsistent with the increase in λ that is expected for wearout mechanisms. The number of failed devices is given by $F(t) = 1 = e^{-\lambda t}$. This distribution can be normalized to $\tau_{63} = 1/\lambda$, where τ_{63} is the value of t where $F(t) = 0.63$.

5.1.2.2 Normal and log-normal distributions

The probability density of the normal distribution is described by

$$f(x) = \frac{1}{\sigma\sqrt{2\pi}} \exp\left[-\frac{(x-\mu)^2}{2\sigma^2}\right] \qquad (5\text{-}5)$$

where μ is the sample mean, and σ is the standard deviation. The normal distribution is usually not applicable to device failure statistics, but a related function, the log-normal distribution is widely used. It is common practice to use a shorthand notation for the normal distribution function

$$\Omega(z) = \frac{1}{\sqrt{2\pi}} \exp-\left(\frac{z^2}{2}\right) \qquad (5\text{-}6)$$

allowing Eq. 5-5 to be written as

$$f(x) = \frac{1}{\sigma}\Omega\left(\frac{x-\mu}{\sigma}\right) \qquad (5\text{-}7)$$

The log-normal distribution can be derived from the normal distribution by assuming that time (in the sense of device life, for our purposes) is related to the variable x in the normal distribution function by the transformation $x = \ln(t)$. The function $f(x)$ can then be transformed into its log-normal equivalent (setting $\mu = 0$ to facilitate the comparison), with the result

$$f(t) = \frac{1}{\sigma t\sqrt{2\pi}}\exp-\frac{1}{2\sigma^2}\ln\left[\frac{t}{\tau_{50}}\right]^2 = \frac{1}{\sigma t}\Omega\left(\frac{\ln(t/\tau_{50})}{\sigma}\right) \qquad (5\text{-}8)$$

where τ_{50} is a normalization time corresponding to 50% failure.

Figure 5-1 shows how $f(t)$, with $\mu = 0$, depends on sigma. For large values of sigma, $f(t)$ is peaked near zero. As sigma decreases, the peak value moves towards $t/\tau_{50} = 1$, producing a higher and higher peak. For small sigma, $f(t)$ becomes sharply peaked, centered at $t/\tau_{50} = 1$.

The number of failed devices for the log-normal distribution is given by

$$F(t) = \Omega\left(\frac{\ln(t/\tau_{50})}{\sigma}\right) \qquad (5\text{-}9)$$

$F(t)$ is plotted in Fig. 5-2. As σ decreases, fewer and fewer failures occur until the value of t is near τ_{50}, consistent with the plot of $f(t)$ in Fig. 5-1.

The median value of $f(t)$ can be used to estimate the mean time to failure, MTF, for the reliability of a population of devices, assuming that the log-normal distribution applies. The result, allowing μ to have values other than zero, is

$$\text{MTF} = \tau_{50} = \exp\left[\mu + \frac{\sigma}{\Omega(0.5)}\right] \qquad (5\text{-}10)$$

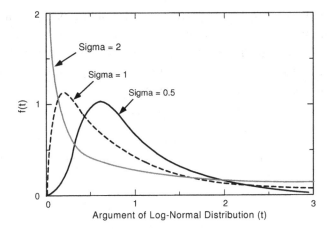

Fig. 5-1. Plot of f(t) for the log-normal distribution for different values of σ. The mean, μ, is set to unity for this comparison.

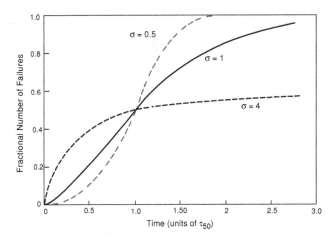

Fig. 5-2. Number of failed devices versus normalized time for the log-normal distribution with μ = 1.

Finally, the failure rate for the log-normal distribution is given by

$$\lambda(t) = \frac{1}{\sigma t} \frac{\Omega\left[\dfrac{\ln(t)-\mu}{\sigma}\right]}{1-\Omega\left[\dfrac{\ln(t)-\mu}{\sigma}\right]} \qquad (5\text{-}11)$$

The shapes of f(t), F(t) and λ(t) vary, depending on the values of σ and μ. Although we usually think of μ and σ in their usual sense of the mean and standard deviation, it is also possible to consider μ as a scale parameter, and σ as a shape parameter for the log-normal distribution. The reasons for this are evident in Figs. 5-1 and 5-2.

5.1.2.3 Weibull distribution

Another distribution function that is often used is the Weibull distribution, given by

$$f(t) = \frac{m}{t}\left[\frac{t}{\tau_{63}}\right]^m \exp-\left[\frac{t}{\tau_{63}}\right]^m \qquad (5\text{-}12)$$

The two parameters that characterize the Weibull distribution are τ_{63}, which is the value of t where the integrated probability reaches 63% of the final value; and m, which describes the "sharpness" of the distribution. Figure 5-3 shows the distribution of f(t). As m increases, the failure density function becomes more sharply peaked.

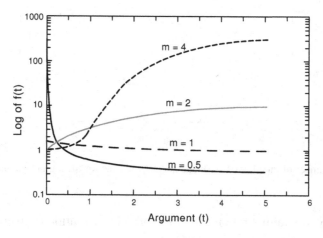

Fig. 5-3. Failure density function for the Weibull distribution.

The number of cumulative failures for the Weibull distrubution is shown in Fig. 5-4 for various values of m.

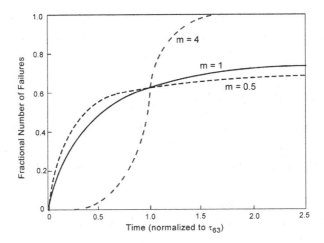

Fig. 5-4. Cumulative failure distribution showing the change in shape for different values of m for the Weibull distribution.

The properties of this distribution are strongly affected by m, just as the properties of the log-normal distribution were affected by σ. With m > 1 the failure rate increases at longer time periods, consistent with the general features of wearout phenomena.

If we compare the log-normal and Weibull distributions with their respective shape parameters set to values that cause the failure density to increase at long time intervals, the Weibull distribution has a failure density that increases less sharply than the log-normal distribution. This results a higher failure rate at short times, but it is only evident if we have a sample size that is large enough to provide a statistically meaningful evaluation of the failure rate in the 0.1 to 1% range.

Typically, reliability evaluations are done using a sample from the overall device population. If a number of tests are done using random samples, then for the case where the number of samples is large the mean value from the *sampling* distribution, $\bar{x}$, will be the same as that of the parent population, with a standard deviation of $\sigma/\sqrt{n}$, where n is the number of devices in the test sample. This is known as the central limit theorem. The standard deviation is somewhat larger if n < 30.

For sample sizes of 50 or more, it is possible to determine whether the log normal or Weibull distribution applies to a set of data by plotting the results using scales that result in a linear plot (this is discussed in more detail in Section 5.1.3 below). A more sophisticated method that can be used with sample sizes as small as 10 is described by Cain [1].

5.1.3 Statistical plotting methods

It is common practice in device reliability to determine whether failure distributions fit a log-normal distribution by plotting the failure rate on axes that provide straight lines for an ideal log-normal distribution. MTF ($t_{0.5}$) and the sample mean, μ, are easily obtained from such failure-rate plots, provided that the measured data fit a straight line. Deviations from linearity may indicate the presence of different failure mechanisms, or that a different failure rate function is required.

Experimental uncertainties may also affect the linearity of the failure distribution. Figure 5-5 shows an example. The data are plotted by using a value that corresponds to the inverse value for the normal distribution function, shown by the scale at the right. The scale at the left shows failure percentages for ease in interpretation.

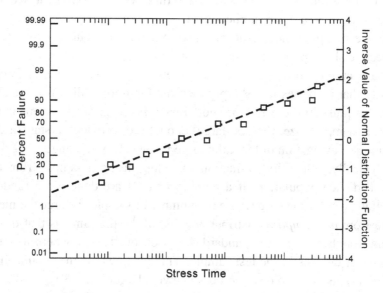

Fig. 5-5. Example of a plot using a vertical axis that results in a straight line for a log-normal distribution.

A similar approach can be used for the Weibull distribution. As discussed earlier, there are significantly more failures for the Weibull distribution than for the log-normal distribution for small values of time, and it may provide a better fit to infant mortality failures compared to the log-normal distribution. Figure 5-6 compares the vertical axes for the special scales that are used for log-normal and Weibull plots.

Note the much wider spacing between the scales for low probability values. Despite these differences, it is difficult to distinguish between fits to the two distributions unless large sample sizes are used.

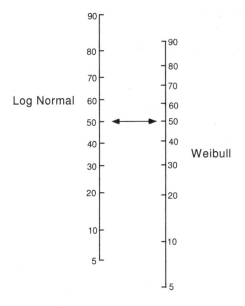

Fig. 5-6. Comparison of the vertical axes for "log normal" and "Weibull" probability paper where data that fit the distribution will lie on a straight line.

5.1.4 Reliability metrics: FIT rate

Device reliability is usually measured by the term *FIT*, which is the number of failures that occur after 10^9 operating hours. One FIT corresponds to a mean failure lifetime of about 120,000 years; an extremely low failure rate for an individual component. Manufacturers typically specify failure rates in the range of 50–1000 FITS, depending on the device technology as well as the number of total devices that are required for specific applications.

As discussed earlier, it is generally impractical to measure such low failure rates under typical operating conditions, and therefore reliability tests are almost always done under highly stressed conditions that increase the failure rate to the point where reliability evaluations can be done in relatively short time periods. Note that the FIT rates for components used in spacecraft are usually lower than required in commercial applications, typically in the range of 1–10. This improved reliability is achieved by conservative derating along with additional testing and screening methods.

For production devices, it is common practice to perform periodic reliability evaluations to determine whether the failure rate is within acceptable limits. Such tests also provide the manufacturer with information on manufacturing variations and trends, which are important in order to produce a reliable product. Devices that are intended for high-reliability applications such as space systems are often subjected to special reliability tests that are done for each production lot, including high-temperature burn in of all devices.

5.2 Acceleration Mechanisms

5.2.1 Temperature: activation energy

The majority of reliability mechanisms are strongly temperature dependent [3]. The failure rate usually increases with temperature, although other factors such as operating current density or applied voltage are even more important for some mechanisms.

It is also possible to evaluate device reliability by examining operational failures at actual working temperature. However, this is rarely done because it requires failure statistics on very large numbers of devices over long time periods, as individual devices have a low failure rate.

Nearly all reliability work is based on accelerated test methods, where a small population of devices operate under elevated stress over a relatively short time interval; periods of days, weeks or months. The failure rate of the sample used for accelerated tests can be applied to devices that operate under normal conditions, based on the assumed temperature dependence for the failure rate (as well as other acceleration factors).

The temperature dependence of thermally activated processes can often be described by the Arrhenius equation

$$R = A e^{-E_a/kT} \tag{5-13}$$

where R = failure rate, A is a constant of proportionality, E_a is the activation energy, which is constant for a specific failure mechanism, k is the Boltzmann constant (8.62 x 10^{-5} eV/K), and T is the temperature in Kelvin. Typical activation energies range from 0.3 to 1.5 eV.

Failure mechanisms that can be described with high activation energies have a stronger temperature dependence than those with low activation energies.

The temperature dependence of a particular failure mechanism can be determined from the Arrhenius equation, yielding the relationship

$$\ln \frac{t_2}{t_1} = \frac{E_a}{k}\left[\frac{1}{T_1} - \frac{1}{T_2}\right] \tag{5-14}$$

where t_1 and t_2 are the measured time to failure based on experimental results for two different sample populations that are stressed at temperatures T_1 and T_2.

The quantity on the right side of Eq. 5-14 can be taken as an acceleration factor, showing the relative increase in failure rate when the temperature T_2 is greater than T_1.

Figure 5-7 shows how the acceleration factor varies with temperature for two different activation energies. Failure mechanisms with small activation energies require higher temperatures in order to increase their failure rate by the many orders of magnitude required for practical reliability studies. This is an important constraint when investigating mechanisms with low activation energies. If other mechanisms such as contact or metallization failure, have more of an effect at higher temperature, the failure rate will increase, interfering with the intended purpose of the elevated temperature test to evaluate the original mechanism.

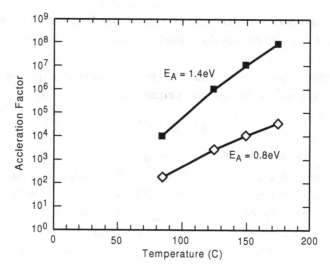

Fig. 5-7. Acceleration factor vs. temperature for two different activation energies.

5.2.2 Infant mortality and burn-in

Real devices have many different failure modes. In many cases the failure rate is not constant with time, but decreases rapidly after a short initial operating period because of "infant mortality". Infant mortality is assumed to be caused by a small number of devices with atypical behavior, arising because of various manufacturing defects that are not present in most of the devices in the parent population.

Once the early failed devices are removed from the population the remaining devices will often exhibit a nearly constant failure rate for extended time periods. As the reliability test proceeds, the failure rate will gradually increase at longer times due to wearout effects. This is shown pictorially in somewhat idealized fashion in Fig. 5-8.

Note however that not all device populations have such a clear distinction between infant mortality and wearout. Devices that operate at very high current (or at high optical power densities for optoelectronic devices) may exhibit wearout-like behavior even after short periods of time.

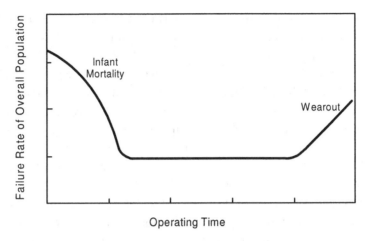

Fig. 5-8. Failure rate vs. time for a population where the failure rate is much higher initially and also increases after a long time due to wearout.

In cases where failures due to infant mortality dominate, it is possible to identify and remove devices with a high early failure rate through burn-in. Burn-in is conducted by operating the entire population of devices for an extended time period under specific operating conditions at elevated temperature (or under other stress conditions, such as high current or voltage) in order to accelerate failure mechanisms. This allows the tests to be done during reasonable time periods, typically days or weeks.

The underlying assumption is that the accelerated wearout will cause weak devices in the population to fail, leaving only the more robust devices with negligible impact on their reliability. The surviving devices can then be used at the much lower temperature and electrical stress conditions involved in normal operation with improved reliability, provided the assumed temperature dependence for the failure mechanism applies and that the acceleration factor is sufficiently high [4].

It is important that the burn-in temperature is low enough to avoid compromising the reliability of the surviving devices. There are various ways to do this, including limiting the junction temperature below acceptable limits for a specific technology. Manufacturers often provide maximum allowable operating temperatures in their specifications.

The effectiveness of burn-in depends on how well the failure rate of the device population is described by the assumed failure distribution, the temperature dependence, and other applicable acceleration factors. Burn-in is not always effective. The temperature must be high enough to cull devices with infant mortality, which may not be possible for mechanisms with low activation energies. Another way that burn-in can be misleading is for populations with non-uniform failure rates. For example, a mixed population from different manufacturing runs may have two or more failure distributions, with different failure rates that may invalidate the assumption of infant mortality.

Failed devices are often discarded, without further examination. It is possible to do a more complete evaluation of failed devices that can, within limits, determine whether the assumed failure distribution is really applicable to the population. This usually requires modifying the burn-in procedure to include additional information. For example, tests can be made at several different intervals during the burn-in cycle, or for two or more groups that are stressed differently. Multiple failure modes, or the presence of devices with distinctly different failure statistics, will usually be evident by evaluating the statistics of device failure. However, this is only possible if the number of failed devices is sufficiently high, and the acceleration factor is the same for all groups.

5.2.3 Other acceleration factors

Other acceleration mechanisms besides temperature can be important. Temperature alone may not be effective for some failure mechanisms, particularly those that involve high voltage or high power density. One way to determine the appropriate acceleration factor is to measure the activation energy of separate samples under different stress conditions. The effect of the stress conditions on the resulting failure rate can be used to select appropriate stress conditions for accelerated tests. Some examples are shown below.

5.2.3.1 Electric field

The electric field is important for mechanisms that depend on hot carriers, impact ionization and specific mechanisms for III-V devices, such as current collapse. An acceleration factor must be developed for the electric field, which depends on the specific mechanism and technology. An exponential dependence, such as that of Eq. 5-15, is sometimes used

$$A = \exp\left(\frac{E_2 - E_1}{\beta}\right) \qquad (5\text{-}15)$$

where A is the acceleration factor, E_2 and E_1 are the two electric fields, and β is the rate constant for the specific acceleration factor and mechanism. The rate factor is not necessarily constant, but may depend on other factors, such as temperature.

5.2.3.2 Current density

Acceleration factors can also be developed for current density. Power-law relationships have been developed that can describe failure rates for metallization or contact failures. However, some mechanisms become important only when the internal current density exceeds threshold conditions [5].

In Chapter 6 we will look at an example for sudden gain degradation in GaAs HBTs where the failure rate increases as the square of current density to the point where the current density dominates the failure rate at high currents, despite the fact that there is also a temperature dependence. Failures in metallization or contacts may also be strongly affected by current density. Any reliability test must consider the fundamental way that variables such as current or voltage affect the failure rate.

5.2.3.3 Non-constant failure rate

The Arrhenius equation is sometimes assumed to describe failure rates, without sufficient supporting information. Figure 5-9 shows an example for PHEMT failures where the activation energy was constant when tests samples were evaluated over a lower temperature range (40–90°C) [2].

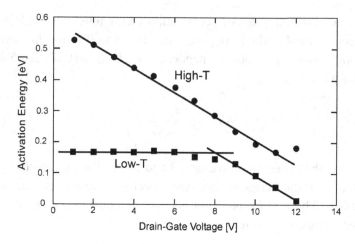

Fig. 5-9. PHEMT activation energy for gate current for various drain-gate voltages [2]. © 2004 IEEE. Reprinted with permission.

However, failure rate tests at higher temperatures (110–160°C) showed that the activation energy changed with voltage, demonstrating that other factors were also important in determining overall reliability under more extreme conditions. This shows how important it is to study device reliability over the entire range of stress conditions that are anticipated for the device before deciding on specific test conditions. In this example, tests at a single high temperature would predict failure rates at reduced (working) temperatures that were too low because the activation energy at more moderate temperatures is so much lower (recall that a high activation energy increases the temperature-acceleration factor).

5.3 Basic Failure Mechanisms

Although failure distributions are useful, they do not provide insight into the underlying mechanisms that cause device failure, and may prove misleading unless they are selected with a clear understanding of the failure mechanisms and their dependencies on temperature and electrical stress. Chapter 6 describes mechanisms for specific devices, including numerous examples. However, before addressing them it is helpful to examine some of the microscopic failure mechanisms and characteristics that can cause failure in compound semiconductors.

5.3.1 Traps at surfaces

Carriers with excess thermal energy can overcome potential barriers within a structure, producing charge in unwanted regions. The excess charge can produce traps at surfaces or at interface regions, which can alter the electrical characteristics of the device [6]. The traps may be stable, increasing gradually under extended operation, or they may gradually recover during operation as bias conditions change.

Trap formation is an important failure mechanism for many compound semiconductors. One way to deal with it is to modify the device structure by adding buffer layers or passivating regions that reduce charge trapping. That approach has been widely used to improve device reliability, but it requires fundamental changes in device design and processing.

5.3.2 Dislocations

Several types of crystal dislocations can occur that affect reliability. Basic categories of dislocations include point, line and plane defects. In an extended crystal lattice more complex dislocations are formed. Stacking faults occur when a line defect alters the crystal uniformity in subsequent atomic layers. Dislocation loops are formed when dislocation lines branch into other dislocations. Threading dislocations are long-range defects in the crystal lattice that are particularly important in active devices.

Dislocations are inherently unstable, and can migrate over time, in some cases forming more complex dislocation regions. Dislocation movement is affected by temperature and operating current within the device. Failure modes associated with dislocations are particularly important for semiconductor lasers.

The dislocation density is generally higher in compound semiconductors than in semiconductors that involve only a single atomic species (*i.e.*, silicon). Some materials, particularly SiC, are very difficult to grow without high numbers of dislocations, which cause reliability in that material to be dominated by crystal imperfections. Dislocations also tend to be higher in materials that are grown on substrates with poor lattice matching, such as GaN on SiC substrates.

5.3.3 Contact degradation

Ohmic contacts are difficult to form in compound semiconductors because of the high barrier height formed between typical metals (Au, Ti, and Pt). They can, however, be formed by creating an alloy, usually by sintering at high temperature.

Schottky electrodes are used in many compound semiconductors. They generally form non-ohmic contacts, but can exhibit ohmic behavior through tunneling when the doping concentration is high.

Both types of contacts can degrade over time because of instability at the interface [7]. Schottky contacts are generally more stable because the interface is inert, but ohmic contacts that depend on alloy formation are more prone to failure because the alloy actually consists of clusters with different chemical composition due to the presence of two elements in the underlying material. The different clusters can grow at different rates during operation at high current density, modifying the interface and increasing contact resistance.

5.3.4 Electromigration

Electromigration occurs because there is an exchange of momentum between electrons, as they move through a conductor or semiconductor, and atoms within the crystal. In a metal region, it causes a gradual drift of atoms towards the direction of the conductor with positive voltage. "Hillocks" can often be seen in metallization after electromigration has taken place. Voids in the structure migrate in the opposite direction, towards the edge with negative voltage.

For thin metallization strips, the result is a gradual "thinning" of the conductor over extended time periods. Cracks eventually occur, causing an open circuit, as shown in the diagram of Fig. 5-10. Regions that are thinner or narrower are more prone to electromigration.

Reliability Fundamentals

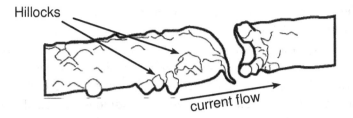

Fig. 5-10. Pictorial diagram showing electromigration in an aluminum metallization line. The metal gradually migrates toward the positive part of the conductor.

Breaks in metal lines often occur where metal lines cross non-planar regions (steps) because the metal is thinner. For aluminum, the approximate threshold where electromigration becomes important is a current density of 10^5 A/cm^2.

Although electromigration has traditionally been associated with metallization, it can also takes place in contact regions. It is an important failure mechanism for MESFETs and laser diodes. For electromigration failures, the mean time to failure, MTF, can usually be described by a power law:

$$MTF = A\,J^n\,e^{-E_a/kT} \qquad (5\text{-}16)$$

where A is a constant of proportionality, J is the current density, n is an exponent (typically between 1 and 2), E_a is the activation energy, k is Boltzmann's constant, and T is absolute temperature.

5.4 Analysis of Reliability Test Data

5.4.1 Screening and infant mortality

Some of the mechanisms involved in failure distributions occur because of abnormal devices that have manufacturing defects. Examples include cracks or dislocations in the starting material, misalignment of wafer masks, and the presence of particles or active impurities that affect processing of a single device.

In many cases, abnormal devices will fail after a relatively short operating period, and, as we have discussed, can be "weeded out" by operating devices for a period of days or weeks at high temperatures under active bias ("burn in").

Although burn in is not used for all technologies, it is generally required for devices used in high reliability applications, particularly for spacecraft [8]. It is an accepted technique for improving the reliability of a group of devices, even if the burn-in conditions exceed the normal envelope for real-use conditions. In order to be successful, the stress conditions must be consistent with the failure mechanisms that are important for the particular technology.

For example, charge trapping is strongly dependent on the electric field, and conditions for burn-in for that mechanism should be designed to apply the maximum electric field to the regions of a device where charge trapping effects will occur. Otherwise the burn-in will likely be ineffective, and will not screen out devices that are highly sensitive to that particular effect. Surviving burn-in does not necessarily produce the desired end result unless the burn-in conditions are carefully planned and supported by more fundamental reliability studies for the particular device technology and process.

It is also important to know – or at least estimate by calculation – the junction temperature during burn-in. If it is too high, the burn-in process may introduce additional failures during normal applications after the burn-in is completed, actually degrading the overall reliability of the surviving devices rather than improving it. This is particularly important for compound semiconductors that operate at high power, or that have substrate materials with poor thermal conductivity.

Another consideration is that several failure mechanisms are potentially important for most devices. The specific conditions for burn in may require tradeoffs, weighting them according to the relative importance of the different reliability failure mechanisms.

5.4.2 Activation energies

Although activation energies do not provide much insight into specific mechanisms, they are widely used to evaluate reliability data. Many failure mechanisms have activation energies between 0.7 and 1.6 eV, and reliability studies often compare the activation energy that is determined from a specific study to activation energies obtained in other reliability work on similar devices. Although such information is useful, it is important not to make too much of it. A similar activation energy does not guarantee that the same mechanism is involved, particularly for devices with very complex structures and processing.

Semi-logarithmic plots of failure distributions can also be extremely useful, showing how well the distribution fits the Arhennius equation as well as how many samples are involved. The earlier example in Fig. 5-9 shows how the interpretation of reliability tests can reveal situations where the activation energy changes with other stress conditions. This is extremely important for compound semiconductors because the structures and processing are very complex, with less mature material technology. This makes it more difficult to establish fundamental failure mechanisms compare to older technologies, *e.g.*, silicon.

Despite the limitations involved in interpreting them, it is still useful to compare activation energies for various failure mechanisms. Typical activation energies are listed in Table 5-1. Many are from studies of silicon technology. Note that there is a considerable range in activation energies even for a specific mechanism. The negative activation energy for hot-carrier degradation is due to the negative temperature coefficient of impact ionization energy.

Table 5-1. Representative Activation Energies for Several Failure Mechanisms.

MECHANISM	TECHNOLOGY	ACTIVATION ENERGY (EV)
ELECTROMIGRATION	AL METALLIZATION	0.5–1
CONTACT FAILURES	AU/TI	0.6–1.1
HOT CARRIERS	SILICON DIOXIDE	-0.06
DIELECTRIC BREAKDOWN	SILICON DIOXIDE	0.3–0.4
DEGRADATION AT HIGH CURRENT	GAAS LEDS	0.9
IONIC CONTAMINATION		0.6–1.4
INTERMETALLIC GROWTH	AU/AL INTERFACE	0.5–2.0

5.4.3 Sample tests

Sample testing is not as straightforward as it appears. The purpose of a sample test is to determine the likelihood that the distribution of values (in this case, the number of parts that fail, or the value of parametric changes) can be bounded by an established limit. This is shown pictorially in Fig. 5-11 for a cumulative failure distribution, assuming a normal distribution function. We cannot answer the question about bounding the distribution in an absolute sense, but it is possible to determine the failure probability from a sample test, subject to confidence limits. Often the sample size used for reliability evaluations is less than 30, requiring a more careful calculation of the statistical uncertainty than the relation $\sigma/\sqrt{n}$ for the standard deviation of the sample mean.

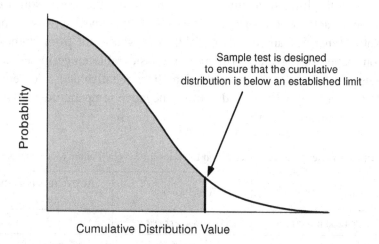

Fig. 5-11. Cumulative failure distribution showing the region where the sample test is intended to bound the parametric change (or distribution of failed devices).

The one-sided tolerance interval approach can be used for the case illustrated in Fig. 5-11, where we are only concerned with values that exceed an agreed upon limit.

The results of the sample test determine the probability that a device will exceed the bounded value, subject to statistical confidence limits. In this context, confidence is used to describe how consistent the sampling results will be if we repeat the sample tests many times. A confidence value of 90% means that we will determine a maximum value for the parent distribution that is at or below the agreed upon limit nine out of every ten times that we repeat the sample test.

To apply this method, the mean value of the test sample, $\bar{x}$, is adjusted by a factor, K_{TL}, that is applied to the standard deviation of the test sample(σ) to determine the upper limit for the parent population:

$$UL = \bar{x} + K_{TL} \sigma \qquad (5\text{-}17)$$

In effect, the small sample size "penalizes" the value of the upper limit because the factor K_{TL} is larger for small sample sizes. Table 5-2 lists value of K_{TL} for various test sample sizes, showing them for three different probabilities, all for a confidence limit of 90%.

The 99% probability at 90% confidence values are approximately the same as three standard deviations for large values of n, but if we have only three devices K_{TL} is about 2 ½ times larger than the "three sigma" value.

Table 5-2. Dependence of the Factor K_{TL} on Sample Size for Various Probabilities at 90% Confidence.

SAMPLE SIZE	99.9% PROB @ 90% CONF	99% PROB @ 90% CONF	95% PROB @ 90% CONF
3	9.65	7.34	5.31
4	7.13	5.44	3.96
5	6.11	4.67	3.40
7	5.20	3.97	2.89
10	4.63	3.53	2.57
20	4.00	3.05	2.21
30	3.79	2.88	1.66

5.5 Summary

This chapter provided a brief overview of the statistical and reliability concepts that are needed for the next two chapters which discuss reliability for specific devices. Temperature increases the failure rate of most mechanisms, and is often used as an acceleration factor for reliability testing. The Arrhenius equation is usually assumed for the temperature dependence of failure mechanisms, although an example was included showing how this can sometimes lead to the wrong conclusion about reliability.

To a large degree the present understanding of semiconductor device reliability is the result of 50 years of experience and evolution of silicon technology. Compound semiconductor devices have a more limited history, and are still evolving, using new material systems that often behave differently compared to silicon. The combination of new failure mechanisms, operation at high electric fields and/or high current density, and immaturity increases the difficulty of dealing with reliability for compound semiconductors. On the other hand, as we are often only dealing with a single operating device rather than large-scale circuits, the overall reliability problem for compound semiconductors can be somewhat easier than for silicon devices.

References

1. S. R. Cain, "Distinguishing between Lognormal and Weibull Distributions", IEEE Trans. on Reliability, **51**(1), pp. 32–38 (2002).
2. R. Menozzi, "Off-State Breakdown of GaAs PHEMTs: Review and New Data", IEEE Trans. on Device and Materials Reliability, (**4**)1, pp. 54–60, 2004.
3. P. Lall, "Tutorial: Temperature as an Input to Microelectronic Reliability Models", 1996 IEEE International Reliability Physics Symposium, pp. 3–9.
4. T. Henderson and M. Tutt, "Screening for Early and Rapid Degradation in GaAs/AlGaAs HBTs", pp. 253–260, 1997 IRPS.
5. J.-S. Rieh, *et al.*, "Reliability of High-Speed SiGe Heterojunction Transistors Under Very High Forward Current Density", IEEE Trans. on Device and Materials Reliability, **3**(2), pp. 31–38 (2003).

6. A. Mazzanti, *et al.*, "Physical Investigation of Trap-Related Effects in Power HFETs and Their Reliability Implications", IEEE Trans. on Device and Materials Reliability, **2**(3), pp. 65–71 (2002).
7. G. Sai Saravanan, *et al.*, "Reliability Studies of AuGe/Ni/Au Ohmic Contacts to MESFETs by Accelerated Aging Tests", IEEE International Workshop on Physics of Semiconductor Devices, pp. 462–465 (2007).
8. W. T. Anderson, J. A. Rousos and J. A. Mittereder, "Life Testing and Failure Analysis of PHEMT MMICs", Proc. of the 2000 GaAs Reliability Workshop, pp. 45–52, Nov. 5, 2000.

Chapter 6

Compound Semiconductor Reliability

Semiconductor device reliability is a complex problem. Although it is relatively straightforward to study individual failure mechanisms using test structures, reliability of working devices depends on many different factors and mechanisms, including manufacturing defects.

For compound semiconductors, crystalline defects are far more important than for silicon because of the inherent difficulty of achieving stoichiometry over extended distances in the lattice. Failure modes for some compound semiconductors are heavily influenced by cracks or dislocations, which are generally of second-order importance for silicon. Consequently the approaches to reliability that have been demonstrated to be effective for silicon technology are not always applicable to compound semiconductors.

Another point to keep in mind is that manufacturing technology tends to evolve in ways that suppress key reliability problems. Thus, technical issues that are widely studied during the years that new device structures and materials are developed often become irrelevant once the technology enters production. This makes it difficult to "track" reliability issues. Reliability problems continually evolve, but in most cases they are eventually suppressed as device structures and manufacturing techniques are modified to deal with them. As devices mature, it is often the second- and third-order failure mechanisms that are the most important.

We will see that charge trapping at surfaces is an important reliability problem for III-V semiconductors as well as for SiC. Bulk traps are also important, particularly for GaAs. Numerous changes in device structure have been made to reduce the effect of traps, including adding extra layers to reduce the influence.of surface traps on device characteristics.

It takes many years to develop new material technologies to the point where reliable devices can be routinely fabricated. On the other hand, SiGe, which is produced in much the same way as conventional silicon, has the legacy of more than forty years of development of silicon-based technology. This has made it much easier to develop advanced devices in that material compared to other compound semiconductor materials.

Crystal structure and binding energy also affect failure mechanisms. The binding energies of SiGe and SiC are very high, as evidenced by the high melting point and high temperatures required for impurity diffusion in those materials. On the other hand, most III-V semiconductors have lower binding energies. Impurity diffusion in those materials can take place at temperatures in the range of 400 to 500°C. Consequently, diffusion-related failure mechanisms are far more important for those types of semiconductors than for SiGe and SiC. For comparison, diffusion in silicon, the standard benchmark for device technology, requires temperatures on the order of 900 to 1000°C.

6.1 MESFETs and HFETs: Mature Technologies

6.1.1 Overview

The main focus of this section is on reliability issues for GaAs and InP based MESFETs. Although reliability is still an active research area for those materials, the main problems for GaAs have been resolved, and commercial devices have been available in that technology for several years that have adequate reliability.

InP is still evolving, but modifications to device design and processing appear to have solved the most important reliability issues for that technology as well.

The discussion of device technology in Chapter 3 showed that MESFETs and HFETs are quite complex, involving many different material layers. Not surprisingly, there are many different factors that affect device reliability. Some failure mechanisms may occur for both GaAs and InP based MESFET technologies, while others are only important for one. We can divide the mechanisms into four basic categories:

- Failures that are thermally activated;
- Failures caused by the presence of other materials or contaminants, introduced during processing or present in packaging materials;
- Failures that are mainly caused by high current or power density;
- Failures related to traps at surfaces or heterostructure barriers.

Various failure mechanisms are discussed in the following subsections.

6.1.2 Gate sinking

HEMT devices typically use several materials between the contact (usually gold) and the gate region. When titanium is used it tends to diffuse into the barrier between the gate and the channel, as depicted in Fig. 6-1 for a GaAs HEMT. This causes a permanent shift in threshold voltage, reducing drain current and transconductance. The diffusion process is strongly temperature dependent, and the mechanism is usually only important for devices that operate at high power where the internal temperature is sufficiently high, and there are thin barrier layers. It can be eliminated by using refractory metals instead of titanium in the gate stack, or by increasing the stack thickness of the gate materials [1].

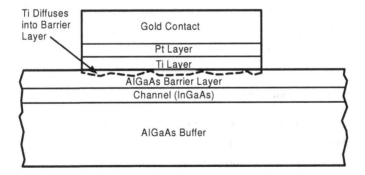

Fig. 6-1. Gate sinking, where titanium in the gate "stack" diffuses into the barrier layer.

This failure mechanism is particularly important for high-power microwave applications that usually involve very high power dissipation. It is a thermally activated process with an activation energy of about 1.5 eV. The time dependence of the diffusion process is proportional to $t^{1/2}$.

Although gate sinking is usually viewed as a reliability problem, it has deliberately been used to improve the performance of InGaP PHEMT devices where it reduces source resistance and increases the Schottky barrier height [2]. The annealing temperature used during processing in that study was 325°C.

6.1.3 Contact degradation

Gold contacts are frequently used for HEMT devices. However, gold diffuses at a relatively high rate through typical barrier metals, forming new chemical compounds that increase contact resistance. Gold diffusion is thermally activated with a typical activation energy of about 1.6 eV. It is particularly important for devices that operate at high temperature, but significant degradation can occur under more moderate conditions as well.

Degradation also occurs for aluminum contacts. The contact resistance increases with the square root of stress time [2], with the same activation energy reported for gold contacts.

Contact degradation can be eliminated by using refractory metals as barriers, or by using improved barrier layers, such as TiN. Most manufacturers have modified their processes to the point that contact degradation is a secondary reliability problem.

6.1.4 Hydrogen poisoning

Hydrogen poisoning is an example of a new failure mechanism for GaAs HEMT devices. It was first reported by reported by Chao, *et al.*, in 1994 [3]. They observed degradation in transconductance and threshold voltage of GaAs as well as in InP HEMTs.

Further work on that topic was summarized by Kayali in 1995 [4]. In GaAs the parametric changes are not necessarily stable. This can cause instabilities in finished assemblies, such as MIMIC amplifiers, and also makes it more difficult to characterize the behavior. Unlike most other mechanisms, hydrogen poisoning occurs relatively rapidly under storage conditions, and does not require active operation to occur.

Hydrogen is introduced during processing. It can also result from outgassing of packaging materials, or other components if the device is in a hybrid package. Concentrations <0.1% can cause measurable shifts in device parameters, and it is possible for hydrogen concentrations as high as 2% to build up in hybrid packages over extended time periods. Gaddi, *et al.*, evaluated the mechanisms for hydrogen-related instabilities by deliberately introducing high concentrations of hydrogen in devices that were passivated with Si_3N_4 [5]. They found that the effect saturated for hydrogen concentrations above approximately 0.1%, and that removing platinum, which is a catalyst for some types of hydrogen-based reactions, did not change the effect (platinum is typically used in gate metallization).

They attributed the change in pinchoff voltage to formation of TiH in the gate, producing strained regions in the semiconductor region with an associated piezoelectric charge. They also proposed that changes in breakdown voltage, which also occurred for their samples, were caused by activation of surface states by hydrogen.

Figure 6-2 shows an example of changes in transconductance and drain current for a device that was subjected to unbiased storage for 90 hours at a temperature of 180°C in a hydrogen-rich environment. Larger changes will occur if the device is operated for longer time intervals.

This particular reliability problem can be overcome by eliminating hydrogen, to the extent possible during processing and packaging, and by including high-temperature storage tests as part of reliability qualification, monitoring device characteristics before and after storage tests.

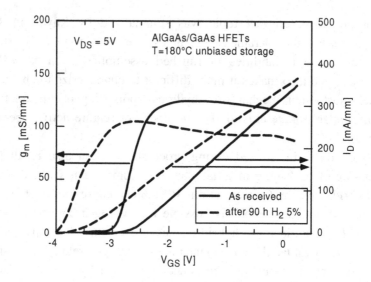

Fig. 6-2. Drift in transconductance and threshold voltage after unbiased storage of AlGaAs HFETs at elevated temperature [5]. © 1999 IEEE. Reprinted with permission.

6.1.5 Fluorine dopant passivation

Fluorine contamination is another mechanism that is affected by small amounts of impurities. One of the undesirable properties of fluorine is that it can passivate dopants in n-InAlAs. There are slight effects in other materials, but the effect is much larger in InAlAs. Fluorine is present in many of the chemicals used during processing and is also present in small amounts in the normal air atmosphere. The main effect of fluorine dopant passivation is to decrease dopant concentration [6]. This leads to several effects in devices including a gradual increase in series resistance and degradation of contacts during extended operation. Because the effect is essentially limited to n-InAlAs it can be eliminated by modifying the device structure to avoid using that material. It is only important for InP-based devices.

6.1.6 Hot-carrier degradation

Carriers with excess thermal energy may have enough energy to overcome the junction barrier height that restricts average carriers within a junction. These hot carriers can become trapped at surface regions in a MESFET structure, changing electrical characteristics. An example is breakdown walkout that may occur in pseudomorphic HEMTs when they are operated near the breakdown voltage limit. It is caused by impact ionization at the edge of the drain, which injects hot electrons into the region between the gate and drain regions [7]. The negative charge buildup that results from this effect causes local widening of the depletion region, which increases the breakdown voltage.

Figure 6-3 shows an example of breakdown walkout after a higher than normal voltage was applied between the gate and drain [8]. That condition often occurs in RF applications. The stress was done by increasing the gate voltage (with the source and drain grounded) until the reverse gate current reached 1 mA/mm. The gate breakdown voltage then increased to a higher value.

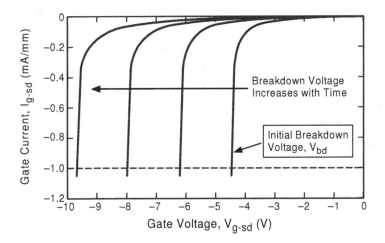

Fig. 6-3. Breakdown walkout after high fields were applied between the gate and drain [8]. © 1992 IEEE. Reprinted with permission.

The curves in the figure show four successive stress conditions, where the gate voltage was increased to the threshold condition for gate current. The breakdown voltage after the last stress condition was stable, even after overnight storage at 150°C.

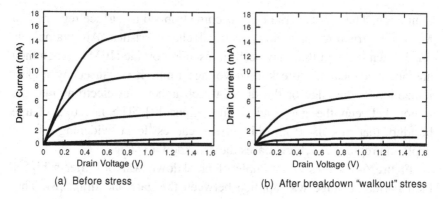

Fig. 6-4. Change in drain current characteristics after excess field stress during breakdown walkout [8]. © 1992 IEEE. Reprinted with permission.

Although an increase in breakdown voltage is usually not an issue, the increase in breakdown voltage caused by hot carriers is usually accompanied by a permanent change in threshold voltage, reducing transconductance. The change in characteristics before and after stress are shown in Fig. 6-4 for an AlGaAs HEMT [8]. In this example, the transconductance has decreased by more than a factor of two after the extended stress condition, a very significant change.

This problem can be solved by modifying the device structure to reduce the electric field in the drain-edge of the gate, or by changing the passivation layer structure. Because it only occurs for high voltage, breakdown walkout can be eliminated by restricting device operation to voltages below the breakdown limit, avoiding the overstress condition. However, this is not always possible because some applications require operation with electric fields that approach – or slightly exceed – the gate breakdown limit.

6.1.7 Passivation layer traps

Several other effects can occur because of charge traps in passivation layers. Mechanisms for charge generation include injection of hot electrons from the channel region or from the region near the drain, where the electric field is the highest. The general effect at the device level is to reduce the maximum drain current (which also reduces output power), and increase drain resistance.

Trapping in passivation layers can be reduced by using different passivation techniques, as well as by modifying the device geometry to reduce the electric field. Several passivation techniques are available, including thin films of Si_3N_4, and the use of thin heterostructures with high doping concentration at the surface. Traps can occur at any of the interface regions that are present in advanced HFET devices [9]. The presence of kinks in the drain current vs. drain voltage characteristics has been attributed to interface traps.

A great deal of effort has gone into modifying device structures to reduce the effects of traps in passivation layers on device performance.

6.1.8 Gate-lag effect

Very slow transients can occur in the drain current when the gate-source voltage is subjected to a rapid voltage change. Although the majority of the drain current switches very rapidly, the fraction of the current with delayed response is effectively removed from the device response during high-speed switching or RF applications. An example of the gate lag effect in an AlGaAs/GaAs HFET is shown in Fig. 6-5 [10] for the transistor structure in Fig. 6-6. (The diagram in Fig. 6-6 shows only one half of the complete structure, splitting it down the line of symmetry). Transient waveforms are shown for devices with different values of gate recess length (ΔL_{g1}), defined in Fig. 6-6, and for two different values of "overdrive" applied to the gate.

Although these devices are capable of extremely high frequency response, the gate-lag effect causes part of the drain current during turn off to extend for several milliseconds, reducing the amplitude of switching pulses. The results in the figure show that the effect is much larger for devices with larger gate recess length. It is also larger when the gate is driven more negative.

Gate lag is attributed to the presence of deep traps at the surface, the semi-insulating substrate, or at heterojunction interface regions. The dependence of gate lag on gate recess length in Fig. 6-5 implies that for this particular device, the traps are located at the recess surface, which has no gate or surface controlling layer.

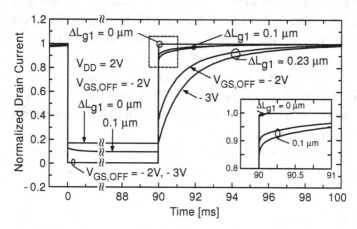

Fig. 6-5. Switching waveform of the drain current for an AlGaAs/GaAs HFET showing the gate-lag effect [10]. © 2003 IEEE. Reprinted with permission.

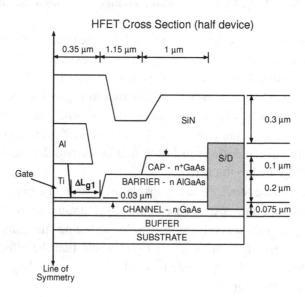

Fig. 6-6. HFET structure used in the results of Fig. 6-5 showing the gate recess length [10]. © 2003 IEEE. Reprinted with permission.

Several factors affect gate lag, including gate voltage, device geometry, and temperature. A high temperature reduces the time constant of the delayed component, but reduces the magnitude of the fast component. The net effect is an increase in the relative size of the slow component at high temperature.

Although reducing the recess extension will reduce gate-lag effects, the voltage breakdown of the HFET will be lower because the impact ionization rate is higher for devices without the recess. Thus, there is a tradeoff between voltage breakdown and gate-lag effects.

Two effects that are closely related to gate lag are power compression (sometimes referred to as power slump), and recoverable power drift [11]. Most high-efficiency amplifiers operate as Class AB amplifiers, driving the gate well below pinch off. Thus, for part of the cycle a high drain voltage is applied with the gate driven below the pinch-off voltage. This introduces a large electric field at the edge of the gate that is near the drain during the switching transition period. Figure 6-7 shows power drift for a GaAs PHEMT during two separate "cycles" of applied power. Note that part of the power drift has recovered during the time that power was removed. Note also that the change in power is relatively small, although the operating time interval was only about 30 minutes, and that larger changes would occur over longer operating periods.

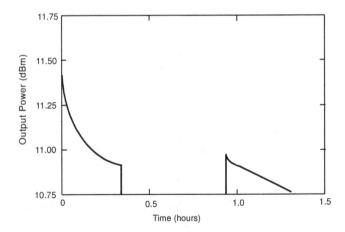

Fig. 6-7. Power drift in a GaAs PHEMT device [11]. © 2000 IEEE. Reprinted with permission.

6.1.9 RF tests of high-power devices for MMIC applications

MESFETs for MMIC applications typically require operation at higher voltages and power levels than MESFETs that are optimized only for high-frequency operation. They are designed to dissipate large amounts of power, and operate under much higher stress conditions. Reliability tests of these devices may be different when tests are done under the high-power RF conditions that represent actual use conditions compared to results with test structures under less severe stress.

Figure 6-8 shows the results of tests done on AlGaAs/InGaAs PHEMTs, with 0.25 µm gate lengths [12]. They were biased with a drain voltage of 7 V, drain current of 450 mA, and an output RF power level of 17 dBm at 8 GHz. The temperature was controlled at the baseplate of the assembly, but the junction temperature of the devices in the assembly was considerably higher. The failure mode observed in these tests was gradual degradation of output power, drain current and the small-signal parameter S_{21}. The failure criterion was a 20% decrease in I_d.

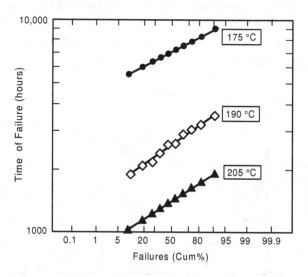

Fig. 6-8. Degradation of MESFETs for MMIC amplifier applications after RF testing for extended time intervals [12]. Note the large decrease in the mean time to failure for the higher temperatures. © 2000 IEEE. Reprinted with permission.

The data fit a log-normal distribution at all three temperatures with similar slopes, justifying the use of the log-normal distribution to evaluate overall reliability. The activation energy was 1.12 V, resulting in an estimated lifetime of more than 10^6 hours at a channel temperature of 140°C. However, it is clear from the data that the temperature rise must be restricted in order to avoid compromising reliability. The mean time to failure is about five times lower at a channel temperature of 205°C compared to the MTF at 175°C.

6.2 GaAs Heterojunction Bipolar Transistors

6.2.1 Basic considerations

GaAs HBTs were developed in the mid 1980s, primarily for use in linear power amplifiers. Earlier devices used beryllium as a dopant in the base region, which has high diffusivity, leading to a temperature-activated failure mechanism that reduced the gain at low emitter-base voltage because the Be creates a conduction-band barrier after it diffuses into the structure. This failure mode could be eliminated by using carbon for base doping, which has lower diffusivity, instead of beryllium. Although SiGe and InP HBTs are gradually supplanting GaAs HBTs, it is instructive to discuss the reliability mechanisms that occur in GaAs devices which are still used in many applications.

6.2.2 Degradation in carbon-doped HBTs

Although carbon essentially eliminates the Be diffusion problem, HBTs with carbon doping exhibit a gradual wearout phenomenon [13]. The effect is particularly severe in structures where there is no passivation layer over the base.

Tests at elevated temperature (238 and 280°C) with fixed base current showed a reduction of about 30% in gain after 500 hours of operation. Much smaller changes occurred in passivated devices.

Electrical tests at low temperature (77 K) after the stress was applied showed a 15% increase in the collector current. They attributed this to a tunneling recombination current from the base to the emitter that was enhanced by surface charge during stress.

For normal temperature and stress conditions this degradation mechanism becomes unimportant when a thin AlGaAs passivation layer is used over the base.

6.2.3 Sudden DC gain degradation

Some types of GaAs-based HBTs exhibit "sudden DC gain degradation" when they are operated at extended time periods with high operating current. Initially the gain degrades slowly in a predictable manner, but after a critical time the rate of degradation increases rapidly, causing the gain to fall to low values over a relatively short time – in some cases only a few hours of incremental stress time - compared to the total stress time.

This is a critical failure mechanism for devices that are operated at high current. Feng, *et al.*, used tests at high temperature under different operating conditions to determine the interplay between temperature and current density on sudden beta degradation [14]. The sample size of each group was 25; the failure criterion was a 50% reduction in gain. The samples fit a log-normal distribution, which was used to determine the mean time to fail under various conditions. The primary focus of that work was on InGaP devices, but they also evaluated AlGaAs HBTs with similar structures for comparison. Figure 6-9 shows how the MTTF varied with current density for the two technologies when they were operated at 305°C.

The slope provides the effective acceleration factor for current density, which is nearly J^2 for AlGaAs and approximately $J^{1/2}$ for InGaP. The much smaller acceleration factor for InGaP is a major advantage for high current applications.

One model that has been proposed to explain sudden beta degradation is the generation of defects in the emitter-base depletion region [15]. The model assumes a two-carrier electron-hole capture process that creates an additional defect after an electron-hole pair is captured. One parameter of the model is a defect-formation energy, E_{RE}, that depends on the material. The modeling approach assumes four different mechanisms for base current: space charge recombination, neutral base region recombination, reverse hole injection, and tunneling currents. The value of n in the Gummel plot is assumed to be 4 for the tunneling component.

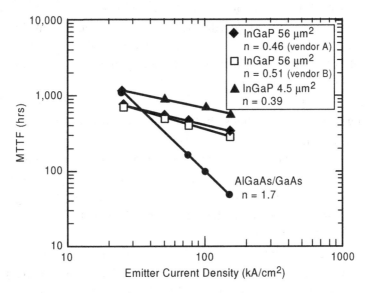

Fig. 6-9. MTTF for sudden beta degradation of InGaP and AlGaAs HBTs [14]. © 2001 IEEE. Reprinted with permission.

Trap densities are assumed to be proportional to space charge recombination and the tunneling component, but are not affected by the other two base current components Simulations with that model predict an increase in base current at low bias conditions, and also a sudden degradation in gain under high bias conditions after an initial period with relatively small change because of the exponential increase in trap density with time.

Although this model cannot explain all of the properties of sudden beta degradation, it does predict that that the REDR rate will be reduced for larger hole barrier heights at the base-emitter heterojunction, which is consistent with the longer mean time to failure of InGaP compared to InGaAs.

6.3 SiGe Heterojunction Bipolar Transistors

As discussed in the introduction to this chapter, SiGe is generally immune to most of the diffusion-related failure mechanisms that occur in III-V devices. Surface trapping is relatively unimportant for two reasons: first, the high quality oxides that exist in the silicon system have far lower surface recombination velocities; and second, the vertical structure of an HBT reduces the effect of surface trapping on device operation compared to the lateral structure of HFETs.

However, there are still important reliability issues. SiGe HBTs have very high cutoff frequencies, but in order to take advantage of this feature in circuit applications it is necessary to operate them at very high current densities. Several mechanisms become important under those conditions including internal heating, electromigration because of the narrow regions, contact degradation, and operation under voltages that are extended into the avalanche region (recall that these devices have very low breakdown voltage).

The increased temperature of SiGe HBTs affects reliability, but degradation under normal or moderately stressed conditions is relatively low. A new technique was suggested by Rieh, *et al.*, to evaluate HBTs under very high current conditions [16]. Figure 6-10 shows the increase in base current for four different stress currents, relative to the normal operating current density of 8.5 mA/μm^2. The tests were carried out at a baseplate temperature of 140°C, resulting in a junction temperature of 192°C for the devices operating at normal current ("1X"). The junction temperatures for the other stress conditions are far higher, up to 372°C for the device stressed at "4X".

Two mechanisms are involved. At short times, the base current increases, which was attributed to an increase in stress-induced traps at the interfacial oxide layer in the emitter. The decrease in I_b at longer times under higher temperature conditions was hypothesized to be caused by passivation of dangling silicon bonds near the poly interface by atomic hydrogen.

The hydrogen is assumed to be released gradually during the stress. The important point of this example is that these devices are extremely robust, even at very high current densities and temperatures well beyond the normal operating range. The use of high current appears to be an effective way to accelerate damage under these conditions, which is a useful reliability diagnostic tool. The authors predict base current shifts of 15% or less under high current operation with normal temperature conditions.

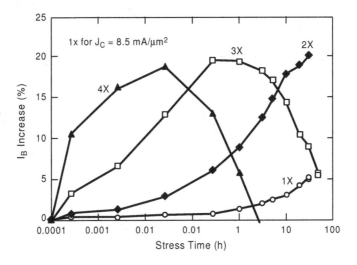

Fig. 6-10. Degradation of an advanced SiGe HBT operating at high current density and elevated temperature [16]. © 2003 IEEE. Reprinted with permission.

The previous work was carried out under voltage conditions that were well within the rated value of collector-emitter voltage. However, better performance can be obtained by operating HBTs at higher voltages, where they are partially in avalanche. Figure 6-11 shows degradation in gain, as well as changes in the Gummel characteristics of a device that was operated at V_{CB} = 3.5 V and V_{BE} = 0.88 V [17]. The normal operating region of this device is at collector-base voltages <2 V, where about 10% of the collector current is due to avalanche multiplication.

After one hour of operation under these conditions there was some degradation at low currents but the gain was essentially unchanged under the high injection conditions that are required for high-frequency operation. The stress conditions in Fig. 6-11 are beyond those recommended for extended operation, and imply that it is possible to operate these devices at high voltage, within the region where significant avalanche multiplication takes place, without compromising reliability.

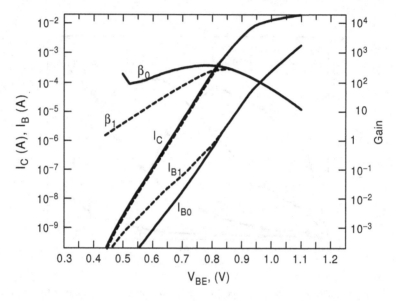

Fig. 6-11. Degradation of an advanced SiGe HBT operating in the forward avalanche region [17]. © 2003 IEEE. Reprinted with permission.

6.4 Wide Bandgap Semiconductors: SiC and GaN

6.4.1 Silicon carbide

A great deal of work has been done on the development of wide-bandgap devices using SiC and GaN. Silicon carbide has been under development for many years. Although the mobility is relatively low, the breakdown field is very high, providing an advantage for high power devices that use high operating voltages. Several commercial devices are available using that material, including bipolar power transistors and high-voltage diodes. As discussed earlier, the main technical problems for SiC are the relatively large number of defects that are present, and the lack of suitable heterostructures. Micropipes in SiC substrates remain a key problem for device reliability, with densities on the order of 10 cm^{-2} for the best material.

Silicon dioxide is the native oxide for SiC. This allows SiC power MOSFETs to be manufactured, which is far more difficult for III-V devices because of the lack of high-quality insulating materials for the gate region. Much of the recent development for SiC has concentrated on power MOSFET technology. One key question is whether suitable oxides can be grown on starting material with higher defects compared to silicon. Figure 6-12 shows lifetime measurements of gate oxides that were grown on two types of SiC surfaces, one the "as-grown" epitaxial layer, which has a higher defect density, and the other an n-implanted surface region on the epi material, with fewer defects [18].

The starting material was 4H-SiC which has a lower defect density than other SiC polytypes. The oxide thickness was 55 to 60 nm.

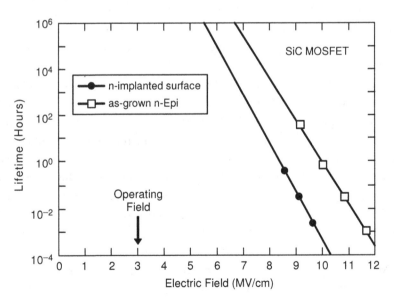

Fig. 6-12. Lifetime measurements of gate oxide capacitors grown on two types of SiC materials [18]. © 2005 IEEE. Reprinted with permission.

Reliability was determined by measuring current through test capacitors as a function of time (oxide leakage current). Both types of structures have extrapolated failure times >100 years at the 3 MV/cm condition corresponding to the maximum gate voltage for finished devices. However, capacitors on the implanted structure have shorter predicted lifetime because of ion implantation damage. This shows that oxides of sufficiently high quality can be grown on SiC, despite the rough surface morphology.

6.4.2 Gallium nitride

Gallium nitride has several performance advantages compared to SiC, and considerable work has been done on reliability in GaN. Heterostructures can be grown using AlGaN (with varying composition), allowing the design of HFET structures, which is not possible for SiC.

Although high-performance GaN devices have been demonstrated at the laboratory level, only a limited number of devices are commercially available. For the most part, GaN devices are still in the developmental stage. Nevertheless, considerable progress has been made for this material technology, and it is likely to evolve rapidly in the near future because of its potential performance advantages in RF applications.

One of the main disadvantages of GaN is the need to grow the material on other types of substrates. Sapphire is often used because the lattice mismatch is smaller compared to SiC, the other substrate material. However, sapphire has much lower thermal conductivity compared to SiC, which is a serious limitation for mainstream applications. The different lattice mismatch and thermal characteristics of the two substrates have to be carefully considered when evaluating GaN reliability, because they can have such a large impact on failure mechanisms. Some work has also been done using silicon substrates for GaN.

One example of a reliability problem is current collapse in GaN HEMTs. This phenomenon is an irreversible change in the I_D-V_{GS} characteristics than occurs when a negative gate voltage is applied that is only slightly below the normal operating range, a condition that is frequently encountered in RF circuit applications. Figure 6-13 shows the results for two different conditions [19]. The top set of curves ("Before stress") show the results of sweeping the gate voltage from -4 to 0 and back to -4 volts for the device. This is within the normal range of voltages for this device. The lower set of curves ("After stress") show the same characteristics after the gate has been stressed by applying a voltage of -5 V to the gate for 10 seconds. The drain current has decreased to 42% of the initial value for this stress condition, which is only slightly beyond the normal range of voltages for linear operation.

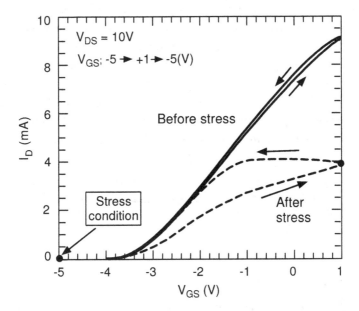

Fig. 6-13. Change in I_D-V_{GS} characteristics of an AlGaN/GaN HEMT after a short-duration negative bias stress (current collapse) [19]. © 2003 IEEE. Reprinted with permission.

Several mechanisms were postulated, including electron injection from deep traps in either the gate barrier level or the interface or buffer layers, or surface states. The effect could be eliminated by passivating the surface with Si_3N_4 suggesting that surface states were the underling cause of the effect.

The high power output that is available in the microwave region is one of the key advantages of GaN HEMTs. Power outputs as high as 30 W/mm have been achieved, but reliability remains an important limitation. Degradation of about 1 dB per 1000 hours of operation is typical of devices that are available at this time. A new passivation method has been proposed that shows a marked improvement in reliability, which consists of pretreatment with NH_3 before the SiN passivation layer is added [20].

Figure 6-14 compares the power output of devices with and without the NH_3 treatment, operated under high power conditions at 2 GHz, with V_{DS} = 20 V and I_D = 200 mA. The off-state breakdown voltage of the devices was 40 V, so the stress conditions are well within the expected range for normal operation. There is a marked difference in the amount of degradation that takes place for devices with different passivation treatments. The improvement was attributed to strengthening of bonds in the material, or possibly the passivation of surface defects from hot electron damage by hydrogen from the NH_3 treatment.

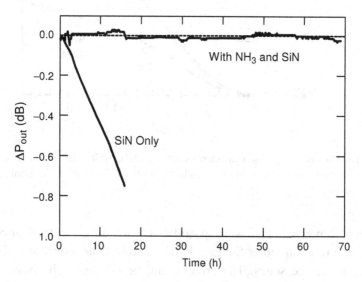

Fig. 6-14. Power output of AlGaN HEMTs with two different passivation treatments [20]. © 2005 IEEE. Reprinted with permission.

Current collapse under RF conditions is still an important limitation for GaN HEMTs, causing much less power to be produced under high-frequency conditions compared to the power at low frequency. Figure 6-15 shows results of a recent study that was done using two different gate dielectric materials [21]. One material was slightly conducting (gate current ~1 μA under normal gate bias conditions), while the conductivity of the other gate was <1 nA.

Two different HFET structures were used, with and without field plates. The field plates overlapped the gate-drain opening by 1 μm. Current collapse was somewhat reduced for devices without the field plate structure when a weakly conducting dielectric was used, but the most dramatic difference occurred for devices with the field plate and conducting dielectric. For that structure, current collapse was eliminated. Power increased linearly up to a drain-source voltage of 55 V, providing essentially the same current available with DC operation.

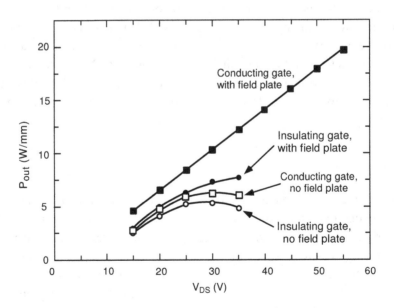

Fig. 6-15. Hot-electron stress results for a GaN HEMT (SiC substrate) [21]. © 2005 IEEE. Reprinted with permission.

A related problem for GaN HEMTs is drain current dispersion due to surface trapping, which causes the transconductance to decrease with frequency. The frequency dependence begins to affect transconductance at approximately 1 kHz, extending to approximately 1 MHz. Menghesso, et al., have developed a model for dispersion and gate lag that takes the polarization charge in GaN HEMTs into account [22].

The energy level of the surface traps was 0.3 eV above the valence band. The model accounts for dispersion and the slow switching response, and implies that the dispersion problem can potentially be solved by modifying the device design with improved passivation methods.

GaN HEMT device technology is still evolving, including extensive work on many different reliability mechanisms. Mainstream applications require high power densities, and significant improvements have been made to understand current collapse and related mechanisms, including charge trapping when the gate is overdriven for RF applications.

The requirement to fabricate GaN on other substrate materials continues to be important. Lattice mismatch between the two different substrate materials makes it difficult to manufacturer reliable devices with satisfactory yield. Reliability evaluations of GaN devices need to consider the effect of such defects as well as conventional mechanisms involving traps and hot electrons in order to achieve satisfactory overall reliability.

6.5 Summary

This chapter discussed mechanisms that affect reliability in compound semiconductors. It is important to have a basic understanding of the underlying reasons for device degradation in order to establish the relationship between application conditions and more basic reliability studies. Some mechanisms are heavily influenced by impurities from processing or packaging, while others are essentially inherent properties of the structure and material technology.

Understanding of reliability issues for GaAs and InP based devices has advanced to the point that it is possible to produce devices with operating lifetime $>10^6$ hours, even under high power conditions. This was done by altering processing and device design, including preparation and passivation of surfaces and by incorporating recessed gates.

In pseudomorphic HEMTs, modifications to channel doping and geometry have been used to reduce the peak fields in the gate-drain region of the channel, reducing hot carrier effects and making devices less vulnerable to current collapse [23].

Work is still progressing on SiC and GaN based devices. Defects are far more important for those materials, which makes them inherently less reliable than the other materials discussed in this chapter. However, significant progress has been made, and devices have been developed for microwave applications that have improved reliability, even under high power conditions, which is the mainstream application where those technologies offer large performance gains.

There are fewer reliability issues for SiGe because the processing is similar to silicon technology, with an extensive heritage that has solved most reliability problems. Note, however, that SiGe HBTs must operate at high current densities in order to achieve the high bandwidth that is required to make that technology competitive with other technologies, which may compromise reliability.

Finally, it is important to keep in mind that most applications of compound semiconductors involve high currents and relatively high power dissipation, even for small-signal applications because the performance improvements that are important for high bandwidth can only be achieved at high current density. The activation energy for many of the reliability mechanisms that we have discussed is above 1 eV, causing the failure rate to be strongly affected by temperature. Reliable operation is critically dependent on keeping the junction temperature below the limits established by more fundamental studies of degradation for the various technologies. This can be far more difficult to do in working devices, unlike test structures, requiring accurate thermal models. Reliability tests must be carefully interpreted. Some of the mechanisms that were discussed are highly sensitive to gate overdrive (which is often required for RF applications). In other cases several mechanisms are involved that may interact with each other, with confusing and contradictory results.

References

1. Y. C. Chou, et al., "On the Investigation of Gate Metal Interdiffusion in GaAs HEMTs", 2003 GaAs IC Symposium, pp. 63–66.
2. C. Canali, et al., "Reliability Aspects of Commercial AlGaAs/GaAs HEMTs", 1991 International Rel. Physics Symp., pp. 206–213.
3. P. C. Chao, M. Y. Kao, K. Nordheden and A. W. Swanson, "HEMT Degradation in Hydrogen Gas", IEEE Elect. Dev. Lett., **15**(5), pp. 151–153 (1994).
4. S. Kayali, "Hydrogen Effects on GaAs Device Reliability", Int. Conf. on GaAs Manufacturing Technology, San Diego, CA, pp. 80–83 (1996).
5. R. Gaddi, et al., "Bulk and Surface Effects of Hydrogen Treatment on Al/Ti-Gate AlGaAs./GaAs Power HFETs", 1999 Int. Reliability Phys. Symp., pp. 110–115.
6. J. A. del Alamo and A. A. Villanueva, "Thermal, Electrical and Environmental Reliability of InP HEMTs and GaAs PHEMTs", 2004 International Elect. Dev. Meeting, pp. 41.1.1–41.1.4.
7. R. Menozzi, P. Cova, C. Canali and F. Fantina, "Breakdown Walkout in Pseudomorphic HEMTs", IEEE Trans. Elect. Dev., **43**(4), pp. 543–546 (1996).
8. P. C. Chao, M. Shur, M. Y. Kao and B. R. Lee, "Breakdown Walkout in AlGaAs/GaAs HFETs", IEEE Trans. Elect. Dev., **39**(3), pp. 738–740 (1992).
9. A. Maazzanti, et al., "Physical Investigation of Trap-Related Effects in Power HFETs and Their Reliability Implications", IEEE Trans. on Device and Materials Reliability, **2**(3), pp. 65–71 (2002).
10. G. Verzellesi, et al., "Experimental and Numerical Assessment of Gate-Lag Phenomena in AlGaAs-GaAs Heterostructures Field-Effect Transistors (FETs)", IEEE Trans. Elect. Dev., **50**(8), pp. 1733–1740 (2003).
11. R. Leoni, III, et al., "Mechanisms for Recoverable Power Drift in PHEMTs", IEEE Trans. Elect. Dev., **47**(3), pp. 498–506 (2000).
12. W. T. Anderson, J. A. Rousos and J. A. Mittereder, "Life Testing and Failure Analysis of PHEMT MMICs", Proc. of the 2000 GaAs Reliability Workshop, pp. 45–52, Nov. 5, 2000.
13. T. Henderson, et al., "Characterization of Bias-Stressed Carbon-Doped GaAs/AlGaAs Power Heterojunction Bipolar Transistors", 1994 IEDM Technical Digest, pp. 187–189.
14. K. T. Feng, L. Runshing, P. Canfield and W. Sun, "Reliability of InGaAsP/GaAs HBTs under High Current Acceleration", 2001 GaAs Symposium Digest, pp. 273–276.

15. R. E. Welser and P. M. DeLuca, "Exploring Physical Mechanisms for Sudden Beta Degradation in GaAs-Based HBTs", 2001 GaAs Reliability Workshop, pp. 135–137.
16. J.-S. Rieh, et al., "Reliability of High-Speed SiGe Heterojunction Transistors Under Very High Forward Current Density", IEEE Trans. on Device and Materials Reliability, **3**(2), pp. 31–38.
17. Z. Yang, F. Guarin, E. Hostetter and G. Freeman, "Avalanche Current Induced Hot Carrier Degradation in 200 GHz SiGe Heterojunction Bipolar Transistors", 2003 IEEE International Reliability Physics Symposium, pp. 339–342.
18. S. Krishaswami, et al., "Gate Oxide Reliability of 4H-SiC MOS Devices", 2005 IEEE International Reliability Physics Symposium, pp. 592–593.
19. T. Mizutani, et al., "A Study on Current Collapse in AlGaN/GaN HEMTs Induced by Bias Stress", IEEE Trans. Elect. Dev., **50**(10), pp. 2015–2020 (2003).
20. A. P. Edwards, et al., "Improved Reliability of AlGaN-GaN HEMTs Using an NH_3 Plasma Treatment Prior to SiN Passivation", IEEE Elect. Dev. Lett., **26**(4), pp. 225–227 (2005).
21. A. Koudymov, et al., "Mechanism of Current Collapse Removal in Field-Plated Nitride HFETs", IEEE Elect. Dev. Lett., **26**(10), pp. 704–708 (2005).
22. G. Meneghesso, et al., "Surface-Related Drain Current Dispersion Effects in AlGaN-GaN HEMTs", IEEE Trans. Elect. Dev., **51**(10), pp. 1554–1561 (2004).
23. S. C. Binari, et al., "Trapping Effects and Microwave Power Performance in AlGaN/GaN HEMTs", IEEE Trans. Elect. Dev., **48**(3), pp. 465–471 (2001).

Chapter 7

Optoelectronic Device Reliability

7.1 Basic Considerations

7.1.1 Material properties

Degradation mechanisms for light-emitting diodes and laser diodes are different from the reliability mechanisms that dominate conventional microelectronics. Modern laser diodes operate at very high carrier densities, and also produce intense amounts of optical power that is concentrated within narrow regions. The high internal optical power introduces new failure mechanisms and reliability issues, including facet damage, that have no counterpart in electronic devices. Although light-emitting diodes operate at lower optical power levels than laser diodes, many of the mechanisms that we will consider affect those devices as well.

For optical detectors there is a closer parallel with conventional microelectronic reliability because we do not have to deal with mechanisms for light generation, and the power levels are much lower. Although still important, our discussion of detector reliability will be more abbreviated than that of reliability in optical emitters.

Before discussing reliability, we need to briefly review the different materials that are used in optical emitters, along with the range of wavelengths that can be used. Table 7-1 lists the main materials of interest. Most can be used for either lasers or LEDs, but some are primarily used for only one of those two categories.

Table 7-1. Material Systems Used for Optoelectronic Devices.

Material	Wavelength Range (nm)	Primary Application
AlGaInN	380–530	LEDs
AlGaInP	600–700	Lasers
GaAsP	650–710	LEDs
AlGaAs	630–950	LEDs and Lasers
GaAs	920–950	LEDs
InGaAs	900–1100	Lasers and LEDs
InGaAsP	1000–1550	Lasers

Degradation mechanisms for these material systems can be quite different, and it is useful to provide an overall summary of the material characteristics and dominant degradation mechanisms before describing them in detail in the material that follows. In most cases the wavelength range is determined by the properties of the band structure, but there are exceptions, such as GaAsP, where isotonic transitions allow direct transitions through intermediate states within the bandgap.

7.1.1.1 AlGaAs and GaAs

This is the first material system that was used for LEDs and lasers, with a 30-year development history. The wavelength range is well matched to silicon detectors, and consequently devices fabricated with those materials continue to be widely used. Crystalline defects in AlGaAs and GaAs tend to increase during forward injection, and homogeneous bulk damage is an important mechanism for LEDs. Long-range defects and facet degradation may also be important for lasers made with those materials.

7.1.1.2 InGaAsP and InGaAs

Both materials are far less prone to develop crystalline defects under forward bias, which causes them to be inherently more reliable than

AlGaAs at equivalent current densities. The dominant failure mechanisms for InGaAs and InGaAsP do not involve bulk damage, but are generally associated with contact degradation. Threshold current densities for laser operation are lower compared to AlGaAs, decreasing the importance of catastrophic optical damage. InGaAs lasers can only be formed on strained lattices, but the lattice strain does not appear to affect reliability as long as the strained regions used within the device are less than the critical thickness for the material.

7.1.1.3 AlGaInP

AlGaInP is used for LEDs and lasers in the red region of the visible spectrum. Lasers are a relatively recent development, and less information is available about reliability for that material system. However, it is less prone to bulk defects than AlGaAs.

7.1.1.4 GaAsP

GaAsP has a direct bandgap as long as the mole fraction of As is <0.3 (GaP is an indirect material). It is primarily used for LEDs in the visible region. GaAsP LEDs are more stable than AlGaAs LEDs, with less degradation over extended operating times. $GaAs_{1-x}P_x$ with $x = 0.3$ results in a wavelength of 700 nm, with transitions between bound exciton states One major manufacturer of optocouplers uses that particular wavelength in many of their products because the response time of the LEDs is higher compared to other wavelengths within that material technology.

7.1.1.5 AlGaInN

AlGaInN is a relatively new material. GaN has a much larger crystalline defect density compared to the other materials we are discussing, making it more difficult to fabricate reliable devices. However, special fabrication techniques have been developed for blue LEDs that extend the growth over silicon-dioxide regions at the surface, resulting in a much lower defect density in the active part of the device, increasing reliability [1].

7.1.1.6 InGaN

The InGaN/silicon carbide substrate combination is widely used for LEDs in the visible region. The wavelength (450–535 nm) can be used to produce LEDs in the blue and green region with very high output power. It is also possible to use this material in conjunction with phosphors or other optically activated impurities to produce "white" LEDs. Although InGaN reliability appears to be adequate for applications near room temperature, these devices can fail at high temperature, even during unbiased storage.

7.1.2 Operating conditions and failure definitions

LEDs and laser diodes can be applied with a variety of electrical conditions. They can operate at high current densities, using short-duration pulses, or under steady-state conditions at lower currents, and also over a relatively wide range of temperature. It is not possible to develop a single set of operating conditions and failure criteria that will be applicable to all applications. The accepted approach is to determine the activation energy and appropriate current acceleration factor that allows reliability test data to be extrapolated to other operating conditions. However, the tests must encompass a range of operational conditions that can validate the activation energy and acceleration factors that are used. Typical operating conditions and failure definitions are discussed below.

7.1.2.1 Light-emitting diodes

LED reliability testing is usually done under constant current injection, measuring the output power along with the current-voltage characteristics. The current selected for stress measurements is typically the maximum rated steady-state current. Devices are typically placed on an extended heat sink, controlled to a specific temperature. The junction temperature is higher than the heat sink temperature, and is usually determined by calculation.

For most LEDs, the output power decreases steadily during extended operation when they are operated near maximum current. A 20% decrease in power output is often used as a failure criterion, although other definitions can be used. Unit-to-unit variability can be evaluated by measuring the total number of operating hours on many devices, plotting the failure distribution to determine how well it is described by a lognormal or other applicable failure distribution.

7.1.2.2 Laser diodes

For reliability studies, laser diodes are usually operated under constant optical output power, monitoring the output of individual devices and increasing the drive current as needed to maintain the output power condition. Thus, injection conditions *change* during reliability tests of laser diodes as the devices degrade. The injection conditions will be different for devices within a test group because they have different threshold currents, as well as different degradation rates. The forward current of a device that undergoes degradation at a relatively short time after testing begins will be considerably higher during most of the extended test time than that of a "typical" device from the same group. The higher injection conditions will tend to accelerate failures in the weak device.

Although this is a reasonable way to evaluate devices if it corresponds to actual use conditions, it makes it far more difficult to evaluate failure mechanisms and to make comparisons between different parts within a test group because the stress conditions are not the same.

Failure is often defined as the point where the drive current doubles. At that point there is usually considerable internal heating within the device. The internal temperature may be different for various regions, including the facets as well as the contact region, increasing the difficulty of accounting for the effect of temperature on reliability mechanisms.

Note the difference in the way that lasers are evaluated compared to LEDs. Under conditions where the drive current is increased to compensate for reduced power output, far less degradation can be tolerated for laser diodes because the increased drive current will eventually produce internal conditions within the laser that will compromise reliability. That is not the case for the constant current conditions that are typically used for LED reliability.

7.2 Reliability of Light-Emitting Diodes

7.2.1 General characteristics

For AlGaAs and GaAs LEDs, aging produces a gradual increase in the number of non-radiative recombination centers within the active region. This has two consequences. First, optical power decreases because a larger fraction of the total current goes into non-radiative (N.R.) processes. Second, the increase in N.R. recombination causes the forward voltage to decrease as the device degrades. Figure 7-1 compares the forward current characteristics of an AlGaAs LED before and after operation at maximum current for approximately 1500 hours. The dashed lines show the power output and forward current after aging, while the solid lines show the initial values. In this example the light output has decreased slightly more than 20%.

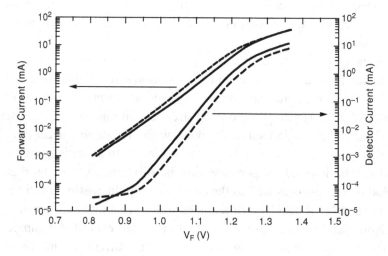

Fig. 7-1. Optical power output vs. forward injection for an AlGaAs LED before and after aging.

In most cases LED degradation is a gradual process, with less concern about sudden catastrophic failure modes compared to lasers [2–3]. Mechanisms for LED degradation include (1) the gradual increase in non-radiative recombination centers discussed above; (2) facet damage, which is associated with oxidation of aluminum in AlGaAs LEDs, but can also take place in other materials; (3) dark-spot defects; and (4) contact degradation. Catastrophic optical damage can also take place in LEDs, particularly for high-bandwidth devices that operate at high power densities because of the thin active region that is required.

An example of degradation of three different types of LEDs in the near infrared region is shown in Fig. 7-2. All three are high-reliability LEDs intended for the aerospace market, with a maximum steady-state current ratings of 100 mA. The data show the mean degradation of ten units of each type, operated at 100 mA at room temperature with a heat sink to maintain the case temperature at 25°C. The 820 nm LED, which is fabricated with a double heterostructure, degraded less than the other two types, which are made with an older amphoterically doped process (see Chapter 4).

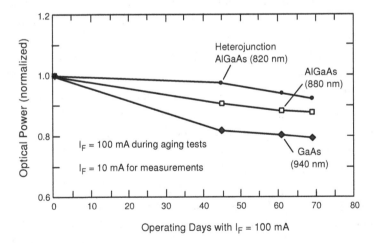

Fig. 7-2. Degradation of AlGaAs and GaAs LEDs after extended operation at maximum rated output power.

Amphoterically doped LEDs require long internal lifetimes, making their characteristics more sensitive to bulk damage compared to heterostructure LEDs that have narrow junctions. These results are for operation at room temperature for a relatively small sample size. The devices would degrade more rapidly if they were used at higher temperature. If more samples were included in the test sample, it is very likely that some of the devices would degrade much more than the typical results shown in the figure, or the worst part in the small test sample. Much larger changes in power output may occur at longer time periods for AlGaAs devices because of dark-line defect formation, which was not a factor in the relatively short duration tests shown in the figure.

Although not shown in Fig. 7-2, samples of the same lot that were operated at 50 mA rather than 100 mA showed far less degradation. Operation at lower currents changes the junction temperature as well as current density, reducing the failure rate. The temperature increase caused by steady state current can be estimated from the thermal resistance. For a metal package with a thermal resistance of 150°C/W, the junction temperature at 100 mA will be about 8°C higher than for 50 mA.

Significantly more degradation occurs in the GaAs LED compared to the other two devices, which use AlGaAs. That is consistent with an earlier observation by Ettenberg, *et al.*, which showed improved reliability for AlGaAs LEDs with an aluminum concentration of 6% or more [4].

Another example of LED aging is shown in Fig. 7-3 for an AlGaAs LED with a wavelength of 660 nm. That composition is near the transition between direct and indirect bandgap for AlGaAs (see Fig. 4-4), which decreases the efficiency because a larger fraction of the recombination current goes into non-radiative processes. The maximum rated current for this device is 40 mA.

If we compare aging results for that device with that of the longer wavelength LEDs in Fig. 7-2, we see that the shorter wavelength LEDs degrade very rapidly during the first few hours of operation when the forward current is 20 mA, which is ½ the maximum rated current. Rapid initial degradation was not seen in the 820 and 880 nm AlGaAs LEDs of Fig. 7-2, which use a material combination that is much further from the indirect bandgap transition region.

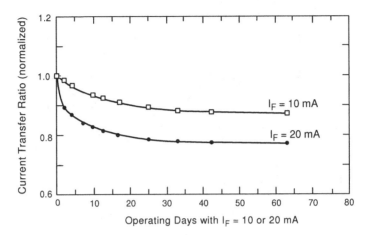

Fig. 7-3. Aging results for 660 nm AlGaAs LEDs from a commercial optocoupler. The maximum steady-state current for these devices is 40 mA. The tests were done at 25°C.

7.2.2 Mechanisms

The above examples show the effect of a gradual build-up in non-radiative recombination centers. Not all LEDs degrade in such a predictable manner. Dark defects can occur when LEDs are operated at high temperature (~80°C or more). They occur abruptly, and are oftem the dominant failure mechanism, particularly for operation at high current.

Figure 7-4, after Fukuda [2], shows degradation of AlGaAsP LEDs that use a contact technology which can potentially create dark spots in the LED (not all units exhibited that behavior). For some units, dark defects developed after about 3000 operating hours, causing a sharp increase in the aging rate. Note, however, that these life tests were done at 150°C, considerably above the normal operating temperature. Dark defects did not develop in these devices when tests were done over extended time periods at room temperature.

The results in Fig. 7-4 show the potential confusion that can occur when tests are done at elevated temperature. In this example the higher temperature introduced a second mechanism that skewed the failure results, because the second mechanism is unimportant at normal application temperatures.

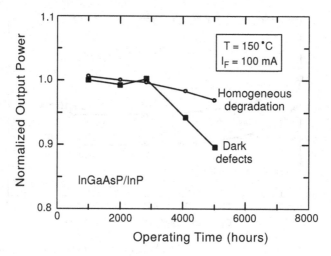

Fig. 7-4. Degradation of two InGaAsP/InP LEDs showing the change in degradation characteristics that is typical of a sample with dark defects along with another unit that exhibits homogeneous degradation [2]. © 1998 IEEE. Reprinted with permission.

Temperature, current density and optical power all affect degradation, and it is not straightforward to apply results taken under one set of conditions to a different operating condition. One way to deal with this is to develop (or apply) simple models for degradation. However, the validity of the models and the underlying assumptions about reliability need to be verified, particularly when more extreme conditions are used in order to accelerate failure rates.

For most LEDs, the rate of change in optical power during extended operation decreases with time. Older studies have often assumed an exponential dependence for the failure rate. However, more recent work has shown that the time-to-failure can be modeled more accurately with a log-normal distribution [5]. The approach recommended by Xie, *et al.*, is to evaluate devices at elevated temperature, leaving the devices in the test until *all* units have failed (they defined failure as a 10% decrease in output power). This allows the entire failure distribution to be determined by plotting the results with a log-normal y-axis. This is a more accurate way to determine failure characteristics compared to tests at room temperature, where only a small fraction of the devices actually fail.

As discussed earlier, dark-spot defects (DSDs) often form during accelerated aging. In addition to degrading output power, they can eventually produce short circuits across the LED junction. Those failures will not fit exponential or log-normal distributions, but often turn out to be the most important failure mechanism because of the abrupt degradation that takes place. DSD failure rates vary widely, and require larger test samples for reliability evaluation compared to gradual degradation mechanisms.

One way in which DSDs can be formed is through migration from the materials in the contact [6]. They compared InP/InGaAsP LEDs with two different contact technologies, BeAu, and Pt. DSDs were only observed in devices with BeAu contacts for the devices in their study. They used secondary ion mass spectrometry (SIMS) to show that the DSDs were formed by diffusion of gold from the contact. The Pt contacts were superior because the diffusion constant for Pt is about two orders of magnitude smaller than for Au.

Dark defects or other non-homogeneous failure modes are usually evident when basic models are used to evaluate reliability because the data corresponding to those defects will depart from the log-normal or exponential distribution. If they occur, they will cause sudden changes in power output that are easy to identify, provided the incremental time steps used for measurement of LED output power are small enough to detect the abrupt changes.

Older LED technologies using AlGaAs or InGaAsP have been developed to the point that homogeneous degradation mechanisms usually dominate the degradation for time periods less than 10,000 hours. However, newer technologies, such as GaN, are less mature, and are prone to sudden catastrophic failure modes.

Even though most LEDs have adequate reliability for operation near their maximum power rating, the usual practice is to limit LED operation to power levels that are at least a factor of two below the maximum steady-state power because of concerns about degradation in highly reliable systems. Most manufacturers include de-rating recommendations in their specifications that are somewhat conservative, and are generally adequate for devices that operate near room temperature. However, it is more difficult to establish acceptable operating conditions for LEDs that operate at elevated temperature.

It is also possible to use LEDs with low duty cycle pulses at power densities that are much higher compared to allowable conditions under steady-state conditions. For example, the heterojunction LEDs in Fig. 7-2 can be used at pulsed currents up to 3 A compared to the 100 mA limit for continuous operation. However, the reliability information provided by the manufacturer usually does not apply to those conditions, and it is up to the user to evaluate how the very high current density under pulsed conditions affects reliability. In many cases the duty cycle for pulsed operation is very low, making it possible to use LEDs under these conditions with satisfactory reliability.

GaN LEDs are less mature, and less is known about the mechanisms that affect their reliability, particularly at higher temperature. A recent study showed that passivated devices degraded far more rapidly compared to devices that were unpassivated, as shown in Fig. 7-5 [7]. The test results in this figure were carried out at 250°C, far above normal operating temperature.

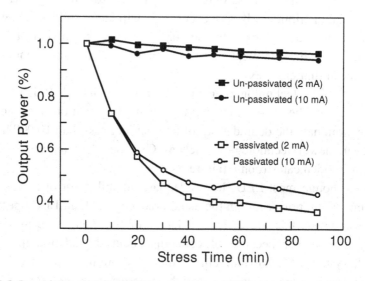

Fig. 7-5. Degradation of InGaN LEDs operating at 250°C showing the influence of the passivation layer [7]. © 2006 IEEE. Reprinted with permission.

Tests at lower temperature showed that the mechanism was thermally activated with an activation energy of 1.3 eV. It was attributed to diffusion of hydrogen from the SiN passivation layer that was used in the process. The activation energy is high enough to decrease the failure rate to sufficiently low values for applications near room temperature, but it becomes important for applications at higher temperature. It could also be important for burn-in, because the higher temperature used for burn in would activate the hydrogen diffusion mechanism and decrease the reliability instead of increasing it.

7.3 Laser Diode Reliability

7.3.1 General characteristics

One of the main effects of aging on laser diodes is a gradual increase in threshold current. In many respects it is similar to the gradual degradation of LEDs under forward injection. Laser characteristics below threshold (where lasers operate as LEDs) show the same behavior with aging exhibited by LEDs. After degradation a higher forward current is needed to overcome those recombination losses in order to produce the internal photon density that is required for stimulated emission, causing an increase in threshold current. The slope efficiency may also change, but that mechanism is usually less important than the increase in threshold current.

Lasers typically have higher electrical power densities than LEDs, along with much higher concentrations of photons along the axis of the active region. Those factors make other mechanisms, several of which result in catastrophic damage, much more important for laser diodes. Some of these mechanisms are discussed in the following material.

7.3.2 Dark-Line Defects

Examination of the emission pattern of lasers after extended operation often shows one or more broad linear regions at the facet where light is no longer emitted. These are called dark-line defects (DLDs), as shown pictorially in Fig. 7-6. They are similar to the dark spot defects discussed for LEDs, but tend to develop far more rapidly because of the high power densities of laser diodes. AlGaAs lasers are particularly prone to DLDs.

In AlGaAs DLDs are formed by the propagation of more localized defects through a process called dislocation climb. One model for this process is dissipation of excess energy at localized defects by the emission of additional point defects that gradually produce a long dislocation dipole along the $\langle 1\ 0\ 0 \rangle$ crystal direction [8]. The excess energy results from the high localized temperature that is produced from non-radiative recombination at the original short-range defect. Once the defect crosses the active region, an extended region is created where non-radiative recombination dominates, producing a dark line that can be observed in the emission pattern.

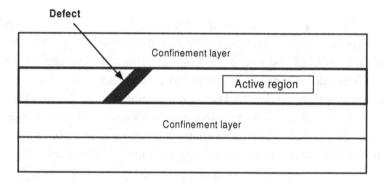

Fig. 7-6. Pictorial example of dark-line defects observed at the facet of a laser diode after aging.

Methods to reduce DLDs depend on the crystal orientation used for fabrication. In $\langle 1\ 0\ 0 \rangle$ AlGaAs material, DLDs can be reduced by using initial substrate materials with very low dislocation density and low oxygen concentration, and the use of "getters" to improve stoichiometry. For $\langle 1\ 1\ 0 \rangle$ material, the key to reducing DLD growth is reduction in mechanical stress after material growth, including the use of soft solders and close matching of the thermal expansion coefficient of the laser material and heat sink. Dark-line defects are much less prevalent in InGaAsP compared to AlGaAs.

7.3.3 Catastrophic optical damage

Laser diodes often exhibit a steady decrease in optical power during continual operation, which is essentially a classic wearout mechanism similar to that of LEDs. However, device failure sometimes takes place very suddenly, with the optical power decreasing rapidly over a much shorter time period. This is called *catastrophic optical damage* (COD). Figure 7-7 shows an example of COD for GaAs laser diodes (860 nm wavelength) operating under constant injection [9], not under the constant power condition that is usually used for laser reliability tests. Note the wide difference in the elapsed time between the start of the reliability test and the time that COD occurs for the different devices in the test sample. These devices were operated near their maximum rated power, 100 mW per facet.

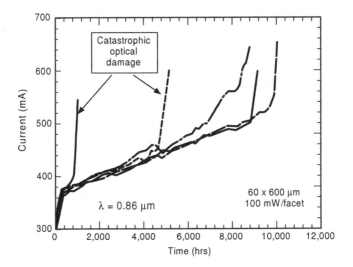

Fig. 7-7. An example of catastrophic optical damage [9]. © 1993 IEEE. Reprinted with permission.

COD occurs when an extended defect is created in the active region of a laser that extends laterally along the cavity length [10]. The effect of this line defect is to increase the lasing threshold, damaging the reflective facet of the laser near the defect region. The surface damage at the facet is the result of localized heating of the semiconductor region near the melting point [11].

Examination of lasers after COD shows that the damaged regions have a higher Al content compared to the surrounding regions, which is consistent with the segregation coefficient of AlGaAs when it is near melting. Another feature of the damaged region is extension of the absorption coefficient to energies that are below the energy gap, indicating that the absorption coefficient is higher in the damaged region. The threshold power level for COD damage in AlGaAs lasers is in the range of 9–20 MW/cm^2.

Catastrophic optical damage can also occur in lasers with other materials, even those that do not contain aluminum. Hashimoto, *et al.*, investigated COD in InGaAs/GaInP lasers, emitting at 980 nm [12]. They fitted their test results for COD to a Weibull distribution, using the 63.2% failure point on the distribution to define the critical power level for COD. The lasers were operated for 100 hours at constant power at a temperature of 50°C to determine the failure statistics.

They also compared the failure statistics of fresh lasers with other groups that had been operated for extended time periods. The critical power level for COD was found to decrease for lasers that operated at longer time intervals. They defined the COD rate, in units of mW/decade of operating time, to show how longer operating times affected the critical power level. The effect of extended operation on operating current was very strong, increasing with approximately the eighth power of operating current, as shown in Fig. 7-8. The current dependence is strong enough so that the probability of COD can be very low, provided the laser is operated well below the current where the COD rate begins to rise so steeply.

Reliable operation for 10^5 hours or more can be achieved with lasers operating below 250 mA, but the steep current dependence causes COD to severely reduce the lifetime for lasers operating at 300 mA. The power level where COD starts to limit the operating life was about 70 mW per facet for the uncoated lasers used in their work, which is in the same approximate range where COD occurs in AlGaAs lasers.

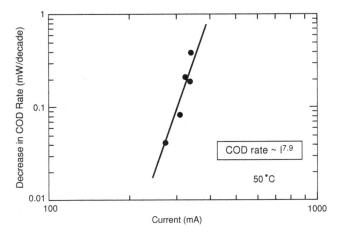

Fig. 7-8. Effect of operating current on the rate of decrease for the critical COD power level in an InGaAs/GaInP laser. The rate increases as approximately the eighth power of operating current [12]. © 1997 IEEE. Reprinted with permission.

7.3.4 Facet damage

Evaluation of degraded lasers shows that high absorption may occur at the facet region after aging. The underlying cause of facet degradation is oxidation of the facet that is enhanced by high light intensity if the photon energy is greater than the bandgap of the facet material. This process, photo-enhanced degradation, has a logarithmic dependence on time for many materials at room temperature.

An example of an experiment to evaluate facet degradation is shown in Fig. 7-9 [13]. InGaAsP and InGaAs lasers were subjected to aging tests at different power levels (constant for each sample in specific test groups). Some of the aged devices were removed after specific time intervals, using argon sputtering to determine the thickness of the oxide that had gradually developed on the facet during aging. Most of the samples were InGaAsP.

Note how much more rapidly the facet oxidation progresses at high power levels. One set of samples was fabricated with InGaAs, and was tested at lower power levels than the InGaAsP samples. However, facet damage still occurred for the InGaAs devices, showing that thicker films develop in InGaAs at much lower power levels compared to InGaAsP.

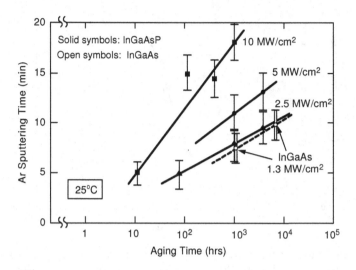

Fig. 7-9. Oxide film thickness after aging for InGaAsP and InGaAs lasers [13]. © 1983 IEEE. Reprinted with permission.

Of the common elements used in compound semiconductors, arsenic has the highest photo-enhanced oxidation rate, and GaAs and AlGaAs are the materials that are most prone to facet damage from this mechanism. Even though arsenic is also present in InGaAsP, the oxidation rate of that material is much lower than for GaAs, reducing the importance of this mechanism compared to AlGaAs. Facet degradation can be reduced by coating the facet with a dielectric film such as Al_2O_3 or SiO_2 that keeps the facet material in the primary material from being exposed to oxygen, as well as by operating the laser at lower power densities.

A recent study was done comparing facet damage in InGaAsP lasers, using photoluminescence to measure the facet temperature with an external laser of lower wavelength [14]. The devices were aged by operating them at a forward current of 2 A. Bulk temperature and facet temperature measurements of devices with and without dielectric film passivation are shown in Fig. 7-10, using a current of one ampere for parametric evaluation after aging. The bulk temperature changed very little, but there was a significant change in facet temperature as the devices degraded during the first 50–60 hours of operation. The passivated samples had smaller changes in temperature after aging. These results show that facet damage is still important, even for lasers that do not contain aluminum. Damage at the facets absorbs some of the photons, increasing the facet temperature even though the bulk temperature of the laser is essentially unaffected.

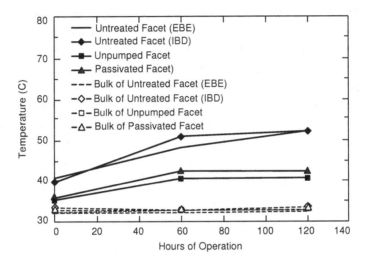

Fig. 7-10. Effect of aging on facet temperature for passivated and unpassivated AlGaAsP laser diodes [14]. © 2005 IEEE. Reprinted with permission.

7.3.5 Electrode damage

Damage in electrodes takes place because of instabilities in the interface between the electrode material and the semiconductor that develop as the device is operated over long time periods. Alloy contacts degrade because of reactions between the contact metal and the semiconductor, particularly when gold is used for the contact. (This is similar to the process discussed earlier in Section 5.3 for contact degradation in HFETs). The reactions produce a non-uniform "spike" region between the contact and the semiconductor that increases with time. Migration of gold can occur at temperatures as low as 200°C in InP [15].

The interface of Schottky electrodes is inert, and therefore is stable with time, making Schottky contacts inherently more reliable than ohmic contacts. However, In or Ga in the metal can gradually diffuse, increasing the contact resistance during extended operating times for device designs that incorporate those materials near contacts.

Electrode failure mechanisms are often the dominant failure mode for InGaAsP lasers because the low defect density in that material makes the other mechanisms relatively unimportant. Other contact-related mechanisms, such as excessive heating due to a high surge current if the device is reverse biased, are also important for that material.

Contact deterioration often has a high activation energy, and may be relatively unimportant at typical device operating temperatures. It can be an important interference for reliability tests if the reliability test temperature is too high. It is possible to overlook the actual failure mechanisms that dominate device degradation under real operating conditions at more moderate temperature when reliability tests are done at elevated temperatures, particularly when contact damage is the dominant failure mechanism.

7.3.6 Reliability evaluation

The previous discussions of specific failure mechanisms for laser diodes pointed out that there is considerable variability in the sensitivity of different samples to the various mechanisms. One way to deal with this is to perform reliability tests on a relatively large number of samples, and determine how well the failure rate is described by the statistical methods discussed in Chapter 5. Figure 7-11, with the Y-axis scaled to produce a straight line for a log-normal distribution, shows experimental results for catastrophic optical damage for ridge lasers (980 nm), using strained InGaAs (AlGaAs). They were operated at 150 mW at room temperature [16]. The results show a mean time to failure of 180 hours, with a standard deviation of 0.46.

Note however that only 9 samples were used in the evaluation, limiting the ability to determine the fit of the observed failures to the assumed distribution. The finite time intervals used for device measurements also influence the shape of the distribution. Although this is adequate for evaluation of general reliability trends, more samples are required to support reliability estimates for production devices.

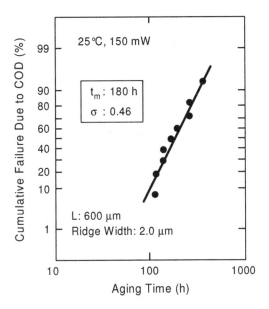

Fig. 7-11. Cumulative failure distribution on the scale for "log normal" paper for catastrophic optical damage in an AlGaAs laser operated at an overstressed power level [16]. © 1994 IEEE. Reprinted with permission.

Despite the small sample size, the aging tests were carried out over a sufficiently long time period that all of the devices failed. This provides better confidence about the fit of the failure rate to the distribution compared to tests where only a small fraction of the test samples are degraded to failure.

A recent example for InGaAs-GaAsP-AlGaAs strain-compensated lasers, operating at 980 nm, is shown in Fig. 7-12 [17]. The typical threshold current was 33 mA. The results in the figure are for eleven devices operating at 40°C at currents that are more than one order of magnitude above the threshold current. Except for one unit (which exhibited an abrupt change of 15% at 1950 hours), output power was essentially unchanged after 3500 hours. Although the results are incomplete because of the finite operating time, they demonstrate the improved performance that modern lasers have achieved during the last decade.

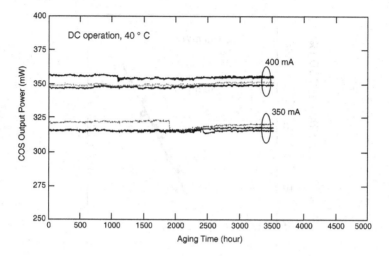

Fig. 7-12. Aging results for 980-nm laser diodes, fabricated with an advance process with low threshold current and high slope efficiency [17]. © 2003 IEEE. Reprinted with permission.

Reliability of modern laser diodes has increased to the point that they can operate for 5000 hours or more before significant degradation occurs. The improved lifetime makes it necessary to do reliability testing under accelerated conditions. A recent example of accelerated aging tests for 1.3-μm laser diodes is shown in Fig. 7-13 [18]. The threshold current of the devices at room temperature was about 20 mA. The reliability tests were done at 150 mA and 85°C, conditions that are much more severe than those encountered during typical operation.

Under these highly stressed conditions the threshold current increased by 15% at 1000 hours, a much shorter test time. Although predictions of operating life assumed that the relative increase in threshold current was linear with aging, the slope actually decreases at longer time periods, adding more conservatism to estimates of operating life.

To validate this approach, other experiments were done to determine the activation energy and the current dependence. Those were combined in the equation below for the mean time to failure, MTF

$$\text{MTF} = A\, J^{-n} \exp\left(\frac{E_a}{kT}\right) \tag{7-1}$$

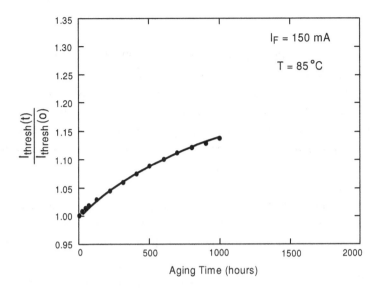

Fig. 7-13. Increase in threshold current for 1.3-μm laser diodes showing a gradual decrease in slope for longer operating times [18]. © 2005 IEEE. Reprinted with permission.

where A is a proportionality constant, J is the current density, n is the exponent for the current dependence, and E_a is the activation energy. The experiments that were done using different current densities fitted this equation with n = 2. The activation energy was 0.55 eV assuming a linear model for the increase in threshold current with aging, and 0.87 eV when a nonlinear model was used to fit the data in Fig. 7-13.

Although Eq. 7-1 is often applied to laser reliability studies the exponent can be very different, depending on the specific design. A reliability study of high-power InGaAs/AlGaAs quantum-well lasers (980 nm) by Yang, *et al.*, found that their devices fit Eq. 7-1 with n = 3.5, a much stronger current dependence [19]. They demonstrated a failure rate of approximately 400 FIT at an operating current of 900 mA. Their lasers interposed a buffer region between the laser cavity and the window to reduce COD. The activation energy for those devices was 0.45 eV. The dominant failure mechanism was random sudden failure. The current used in their reliability tests was about 1/3 of the COD threshold current.

7.4 VCSELs

Only a limited amount of work has been done on VCSEL reliability. One of the complications in dealing with these devices is the wide variation in device geometry. It is possible to fabricate VCSELs with very low threshold current by designing devices with small apertures, with reduced total power output. Although the threshold current is lower for VCSELs with small area, the threshold current density is usually higher than for devices with larger area.

Small-aperture VCSELs have larger losses near the aperture edge. Their maximum operating temperature is usually lower as well. Reliability can be affected by degradation in the distributed Bragg reflectors as well as from non-radiative losses within the laser cavity [20].

VCSEL reliability tests are usually done at constant current, not the constant output power condition that is typically used for edge-emitting lasers. The reason for this is that VCSELs can only operate over a very limited range of currents because of the "rollover" that takes place at higher current due to self-heating.

Initial VCSEL development used AlGaAs technology, operating at 850 nm. Dark-line defects have been observed in AlGaAs VCSELs after aging, along with the gradual degradation that is usually attributed to homogeneous defects for devices using that material. DLDs usually cause catastrophic failure in VSCELs because of the small geometry.

A more complex mechanism has been identified in one type of AlGaAs VCSEL that appears to cause the current path to gradually change during extended operation at junction temperatures above 125°C [21]. Some devices exhibited rapid degradation, while others exhibited the gradual degradation that is characteristic of aging at lower temperature.

Two possible mechanisms were identified, one involving micro-loops from dislocations, and the other migration of point defects, created during aging, to the dopant atoms used to form the mirrors in the VCSEL. The cylindrical geometry will cause more dopant atoms to migrate to the center compared to the edge of the structure, as shown in

Fig. 7-14. The defects increase the resistance at center of the Bragg reflector, causing more current to flow at the periphery of the mirror because of the lower parallel resistance path (recall that the Bragg reflectors are formed by doping atoms). Electroluminscence patterns showed increased light emission from the edge for degraded devices.

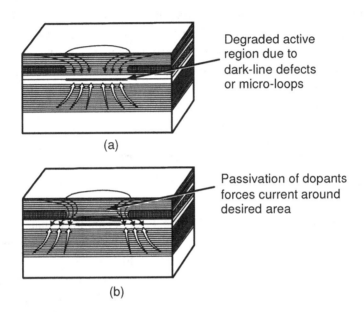

Fig. 7-14. Change in current flow within a VCSEL after aging due to the gradual increase in resistance at the center [21]. © 1998 IEEE. Reprinted with permission.

Although this result does not apply to all VCSELs, it illustrates one way that the different device geometry of a VCSEL can affect the manifestation of a basic failure mechanism – in this case, the gradual introduction of point defects – on device behavior.

There are also many potential applications for VCSELs in the visible range. Recent work on 650-nm VCSELs using AlGaInP has shown larger changes after aging for that technology compared to AlGaAs VCSELs [22]. The estimated operating life of the AlGaInP samples was approximately 1000 hours, compared to more than 10,000 hours for

AlGaAs devices (850 nm). However, the AlGaInP technology is relatively new, and these reliability issues will likely be solved as more work is done on that technology.

Considerable work has also been done on InP-based VCSELs, operating in the wavelength range of 1.3–1.6 μm. Devices have recently been fabricated using selectively etched tunnel junctions to define the aperture [23]. That processing improvement has eliminated the current crowding and scattering losses of earlier designs. VCSELs fabricated in this way achieved differential quantum efficiencies of 25–40%. Although reliability studies have not been completed, the low threshold current and high efficiency suggest that they will have better reliability than older VCSEL structures.

7.5 Tunable and Frequency Stabilized Lasers

7.5.1 Tunable lasers

Tunable lasers represent a more exotic laser technology that is more challenging from the standpoint of reliability because of the need to maintain the tunable characteristics and highly stable wavelength for specific tuning conditions. As discussed in Chapter 4, there are many ways to implement tunable lasers. They generally involve methods that effectively change the length of the laser cavity, or methods that change either the refractive index or select specific resonant modes. Only limited information is available about reliability of tunable lasers, but a recent example is discussed below.

One way to fabricate such lasers is by using additional sections within the laser cavity that affect the "Q". A grating-assisted co-directional coupler tunable laser (GCSR) consisting of a gain section, coupler, a phase adjusting section, and a sampling Bragg reflector, is shown in the simple diagram of Fig. 7-15 [24]. The coarse wavelength setting is adjusted by setting the current though the coupler section, while fine tuning is done by setting current through the phase section.

This particular laser, fabricated with InP, operates at 1585 nm with a tuning range of 7%. The tuning accuracy was ±1 GHz, which is <1 part in 10^5. Current through the coupler is higher at the upper end of the tuning range, which means that the internal power conditions depend on the frequency tuning condition.

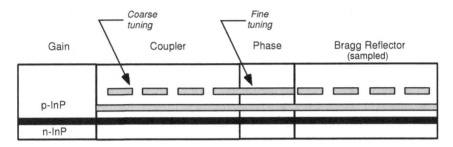

Fig. 7-15. Simplified diagram of a tunable laser using a rear sampled grating reflector and grating-assisted co-directional coupler [24]. © 2000 IEEE. Reprinted with permission.

A reliability study was done by initially burning in the devices for 15 hours at 65°C, removing any devices that failed during burn in [24]. Accelerated aging was done by selecting two groups of 8 samples, operating them continuously at 50 and 65°C. The devices were operated with maximum current through each section of the laser structure, a worst-case condition.

The failure condition was defined as a frequency shift of 10 GHz, which is ten times the accuracy limit that was established prior to aging. A log-normal plot was used to analyze the results, which provided a way to estimate the mean lifetime. The Arrhenius equation was then used to estimate the median lifetime at room temperature. Figure 7-16 shows the results for three different cumulative failure conditions. Based on those results, the room temperature lifetime was estimated to be 13.4 years. This was an experimental laser, not a production device, but the results illustrate some of the issues that are important for the reliability of tunable lasers.

Aging and stress conditions must be tailored to the specific laser design, but generally follow the same principles used for more basic laser structures. Lasers with movable mechanical elements (such as MEMS) will require stress conditions that include potential failure mechanisms associated with those structures.

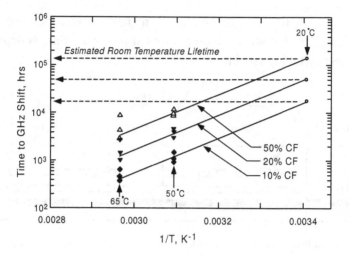

Fig. 7-16. Aging study of a tunable laser for devices operated with maximum current through each section of the laser region [24]. © 2000 IEEE. Reprinted with permission.

7.5.2 Frequency-stabilized lasers

Although we will not provide a specific example for reliability of frequency-stabilized lasers, the previous figure illustrates a case where the initial frequency accuracy was very high, and the failure criterion was a very small change in frequency.

We can address the general issues that affect reliability of stabilized lasers by noting that the linewidth of a single-mode laser depends on the spontaneous emission rate as well as the square of the linewidth enhancement factor that accounts for phase fluctuations within the cavity (Eq. 4-11, repeated below):

$$\Delta \upsilon = \frac{K}{4\pi S}(1+\alpha^2) \qquad (4\text{-}11)$$

Linewidth also depends inversely on the photon density within the cavity. An increase in non-radiative recombination loss would cause a gradual increase in linewidth, as well as mode hopping if the internal conditions degrade to the point where more than one mode can resonate within the cavity. As shown earlier in Fig. 4-21, a narrow laser linewidth is only possible at low optical power levels, and non-radiative recombination losses will compete with single-mode operation in that region. Based on that result, one would expect that lasers with very narrow spectral width would have more difficulty maintaining their specifications over extended operating periods.

7.6 Optical Detectors

7.6.1 Conventional detectors

Relatively few studies have been done to evaluate the reliability of InGaAs photodetectors. Surface leakage currents in InGaAs p-i-n detectors can be reduced by adding a thin InP cap layer above the InGaAs layer, which is the key in making reliable devices. This type of p-i-n detector has very high reliability, with activation energies for failure modes of about 1.5 eV.

Figure 7-17 shows the results of one study where the parent population was divided into three groups, using three different temperatures [25]. Note the large difference in the mean time to failure for samples tested at different temperatures. They extrapolated the lifetime at 70°C to 1.6×10^{10} hours, corresponding to 0.06 FIT, an extremely low failure rate.

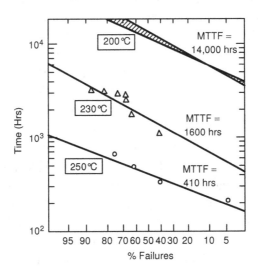

Fig. 7-17. Life tests of InGaAs photodetectors (after Forrest, *et al.* [25]). © 1988 IEEE. Reprinted with permission.

Although this suggests that p-i-n detectors fabricated with InGaAs will have very high reliability, one drawback of this technology is high sensitivity to ESD. It is possible to improve the ESD performance by using floating field rings.

7.6.2 Avalanche photodetectors

Avalanche detectors are the most challenging detector technology from the standpoint of reliability because high field conditions are necessary within the structure to produce avalanche multiplication. Dark current must remain very low in order to meet signal-to-noise requirements. Avalanche detectors using compound semiconductors typically use multiple layers, isolating the surface and periphery with buffer regions and guard rings to minimize leakage current and increase device stability. The structures are quite complicated, and are often considered proprietary, adding to the difficulty of assessing reliability unless it is possible to get the details of the actual design.

Avalanche gain can be adjusted over a limited range, depending on voltage and temperature. The specific conditions used for reliability tests need to be carefully defined, particularly the gain, because tests done under one set of conditions may not necessarily apply to different operating conditions.

An example of reliability work on an avalanche photodetector is shown in Fig. 7-18 for an InAlGaAs superlattice APD [26]. The criterion for failure was an increase in leakage current of 100 nA, with an avalanche gain of ten (that guaranteed that the sensitivity of an optical receiver was degraded by no more than 0.3 dB). A number of samples were used in the study, dividing them into groups that were life tested at three different temperatures. The mean lifetime results are shown in the Arrhenius plot in Fig. 7-18, with an activation energy of 1.05 eV. The extrapolated lifetime at 50°C is approximately 3×10^5 hours.

Most of the devices in the test population showed very little degradation initially, with a relatively abrupt change after operation for several hours (or tens of hours, depending on the temperature).

A more recent study by Yagyu, *et al.*, was done on InAlAs APDs that did not use guard rings [27], eliminating the leakage current mechanism within the guard ring. They identified three different failure modes in their study: (1) increase in dark current; (2) decrease in breakdown voltage, which must be accounted for in applications because of the dependence of APD operation on the avalanche multiplication factor; and (3) a short circuit mode. The short-circuit mode was attributed to metal migration from the electrode, because it occurred at the same time for biased and unbiased samples in the test.

All three mechanisms had activation energies that were high enough to prevent them from appearing in operation under normal operating conditions. The predicted reliability at 85°C was $>10^6$ hours.

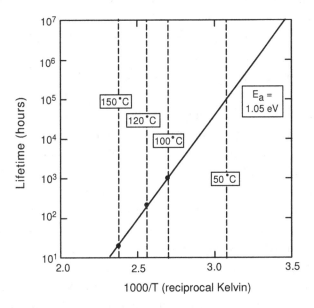

Fig. 7-18. Mean lifetime results for InGaAs APDs tested at various temperatures. The failure criterion was an increase in leakage current of 100 nA [26]. © 1996 IEEE. Reprinted with permission.

7.7 Summary

This chapter has discussed reliability in optoelectronic devices. For light emitting diodes and lasers, the picture is rather complicated because of the very different properties of the materials that are used. AlGaAs tends to develop recombination centers and dislocations over time, while InGaAsP does not. The gradual increase in recombination centers decreases the light output of LEDs, and increases the threshold current of GaAs and AlGAs-based lasers. Although catastrophic optical damage can occur in all lasers, that failure mechanism is particularly important for InGaAsP. It is possible to deal with COD by limiting operating conditions to regions that are well below the threshold for damage from this mechanism.

Studies of devices with different facet passivation have demonstrated the importance of facet oxidation, which is enhanced by high optical power. Facet degradation is one of the key limitations in InP-based lasers. It can be reduced by using coatings to reduce facet heating due to absorption of energy in the facet material.

Many improvements that have been made in fabricating LEDs and laser diodes, which have successfully increased the mean lifetime to periods of thousands of hours or more. Devices with even longer lifetimes have been developed for long-haul communications applications. Several factors have contributed to these reliability improvements, including modifying the structure to reduce threshold current and increase efficiency, decreasing internal operating power and temperature, and improving contact technology to reduce electromigration and diffusion. Although we have not discussed packaging and assembly, reliability also depends on mechanical details and stresses introduced at that level.

Detector reliability is more straightforward because much lower power levels are involved, eliminating the most troublesome mechanisms that affect optical emitters. Detectors with high electric fields – particularly APDs – are affected by surface leakage and changes in dark current. Very complex structures are used for III-V APDs that are designed to limit electric fields and reduce edge and surface leakage. Recent results indicate that the lifetime of these devices can be several years or more, but it important to recognize that reliability in APDs is strongly affected by specific use conditions.

References

1. R. F. Davis, *et al.*, "Gallium Nitride Materials – Progress, Status and Potential Roadblocks," Proc. IEEE, **90**(6), pp. 993–1004 (2002).
2. M. Fukuda, "Laser and LED Reliability Update," IEEE J. of Lightwave Tech., **6**(10), pp. 1488–1494 (1988).
3. O. Pursiainen, *et al.*, "Identification of Aging Mechanisms in the Optical and Electrical Characteristics of Light-Emitting Diodes," Applied Phys. Lett., **79**(18), pp. 2895–2897 (2001).
4. M. Ettenberg, H. Kressel and H. F. Lockwood, "Degradation of Al $_x$Ga$_{1-x}$As Heterojunction Electroluminescent Devices," Appl. Phys. Lett., **25**(1), pp. 82–85 (1974).
5. J. Xie and M. Pecht, "Reliability Prediction Modeling of Semiconductor Light Emitting Device," IEEE Trans. on Device and Materials Reliability, **3**(4), pp. 218–222 (2003).
6. A. K. Chin, *et al.*, "Direct Evidence for the Role of Gold Migration in the Formation of Dark-Spot Defects in 1.3 μm InP/InGaAsP Light-Emitting Diodes," Appl. Phys. Lett., **45**(1), pp. 37–39 (1984).
7. M. Meneghini, *et al.*, "High-Temperature Degradation of GaN LEDs Related to Passivation," IEEE Trans. Elect. Dev., **53**(12), pp. 2981–2987 (2006).
8. S. O'Hara, P. W. Hutchinson and P. S. Dobson, "The Origin of Dislocation Climb During Laser Operation," J. Appl. Phys., **30**(8), pp. 368–371 (1977).
9. S. L. Yellen, "Reliability of GaAs-Based Semiconductor Diode Lasers: 0.6–1.1 μm," IEEE J. of Quant. Elect., **29**(6), pp. 2058–2067 (1993).
10. B. W. Hakki and F. R. Nash, "Catastrophic Failure in GaAs Double-Heterostructure Injection Lasers," J. Appl. Phys., **45**(9), pp. 3907–3912 (1974).
11. C. H. Henry, *et al.*, "Catastrophic Damage of Al$_x$Ga$_{1-x}$As Double Heterostructures," J. Appl. Phys., **50**(5), pp. 3721–3732 (1979).
12. J. Hashimoto, *et al.*, "Aging Time Dependence of Catastrophic Optical Damage Failure of a 0.98-μm GaInAs-GaInP Strained Quantum-Well Laser," IEEE J. Quant. Elect., **33**(1), pp. 66–72 (1997).
13. M. Fukuda, "Facet Oxidation of InGaAsP/InP and InGaAs/InP Lasers," IEEE J. Quant. Elect., **19**(11), pp. 1692–1698 (1983).
14. A. Chavan, *et al.*, "Comparison of Facet Temperature and Degradation of Unpumped and Passivated Facets of Al-Free 940-nm Lasers Using Photoluminescence," IEEE J. Quant. Elect., **41**(5), pp. 630–635 (2005).

15. A. K. Chin, et al., "Direct Evidence for the Role of Gold Migration in the Formation of Dark-Spot Defects in 1.3 µm InP/InGaAsP Light-Emitting Diodes," Appl. Phys. Lett., **45**(1), pp. 37–39 (1984).
16. M. Fukuda, et al., "Degradation Behavior of 0.98 µm Strained Quantum Well InGaAs/AlGaAs Lasers under High Power Operation," IEEE Journal of Quant. Elect., **30**(2), pp. 471–476 (1994).
17. J. Zhao, L. Li, W. Wang and Y. Lu, "High-Power and Low-Divergence 980-nm InGaAs-GaAsP-AlGaAs Strain-Compensated Quantum-Well Diode Laser Grown by MOCVD," IEEE Phot. Tech. Lett., **15**(10), pp. 1507–1509 (2003).
18. J.-S. Huang, "Temperature and Current Dependence of Reliability Degradation of Buried Heterostructure Semiconductor Lasers," IEEE Trans. on Device and Material Reliability, **5**(1), pp. 150–154 (2005).
19. G. Yang, et al., "Highly Reliable High-Power 980-nm Pump Laser," IEEE Phot. Tech. Lett., **16**(11), pp. 2403–2405 (2004).
20. Y. M. Cheng, et al., "Degradation Mechanisms of Vertical Cavity Surface Emitting Lasers," Proceedings of the 1996 IEEE Reliability Physics Symposium, pp. 211–213.
21. R. W. Herrick and P. M. Petroff, "Gradual Degradation in 850 nm Vertical-Cavity Surface-Emitting Lasers," IEEE J. Quant. Elect., **34**(10), pp. 1963–1969 (1998).
22. A. Knigge, et al., "650-nm Vertical-Cavity Surface-Emitting Lasers: Laser Properties and Reliability Investigations," IEEE Phot. Tech. Lett., **14**(10), pp. 1385–1387 (2002).
23. D. Feezell, D. A. Buell and L. A. Coldren, "InP-Based 1.3–1.6 µm VCSELs with Selectively Etched Tunnel-Junction Apertures on a Wavelength Flexible Platform," IEEE Phot. Tech. Lett., **17**(10), pp. 2017–2019 (2005).
24. O. A. Lavrova and D. L. Blumenthakm, "Accelerated Aging Studies of Multi-Section Tunable GCSR Lasers for Dense WDM Applications," IEEE J. Lightwave Tech., **18**(12), pp. 2196–2199 (2000).
25. S. R. Forrest, et al., "Reliability of Vapor-Grown Planar $In_{0.53}Ga_{0.47}As$/InP p-i-n Photodiodes with Very High Failure Activation Energy," IEEE Elect. Dev. Lett., **9**(5), pp. 217–219 (1988).
26. I. Watanabe, et al., "Reliability of Mesa-Structure InAlGaAs-InAlAs Superlattice Avalanche Photodiodes," IEEE Phot. Tech. Lett., **8**(6), pp. 874–876 (1996).
27. E. Yagyu, et al., "Guardring-Free Planar AlInAs Avalanche Photodiodes for 2.5-Gb/s Receivers with High Sensitivity," IEEE Phot. Tech. Lett., **19**(10), pp. 765–767 (2007).

Chapter 8

Radiation Environments

This chapter discusses radiation environments. Particles in space environments can cause both permanent damage and transient effects in electronic and optoelectronic devices. In most cases device damage is caused by the integrated effect of large numbers of individual particles as they pass through the device structure. However, effects from the interaction of only a *single* particle can also be important, particularly for high-energy cosmic ray particles. We need to carefully distinguish between integrated and single-particle environments in the material that follows. The effects of those environments on devices will be discussed in later chapters.

The main emphasis of this chapter is on radiation in the natural space environment. A brief discussion is also included about terrestrial radiation environments. We will consider two general types of space missions: earth-orbiting missions that are strongly affected by radiation in the Earth's trapped radiation belts, as well as cosmic rays and solar flares; and interplanetary (deep-space) missions, where there are no trapped particles, and the only concern is radiation from galactic cosmic rays and solar flares.

8.1 Particle Types

8.1.1 Particles producing permanent damage

The basic particles that we are concerned with are listed in Table 8-1. The first two, protons and electrons, are present in large numbers in the

Earth's trapped radiation belts. Protons are also present in solar flares, which affect most spacecraft except those in Earth orbits with very low inclination (<30° with respect to the equatorial plane). The trapped particles in the Earth's radiation belts have a distribution of particle energies, up to the maximum values shown in the table. The energy ranges of other particles of interest are also shown.

Protons (and/or electrons) are the most numerous particles in space environments. The integrated effect of many interactions of these particles produces (on average) a uniform amount of damage in a semiconductor device, in contrast to single-particle effects, where the effects are highly localized. Electrons and protons produce electron-hole pairs as they lose energy through the ionization process, the dominant energy loss mechanism. The electrons and holes can produce traps and surface states at the interface region between a semiconductor and insulator, damaging the device. This is referred to as ionization damage or total dose damage.

Muons are also present in space, but their effects on electronics are far less important, and will not be considered here.

Table 8-1. Particles that Produce Permanent Damage.

PARTICLE	ENERGY RANGE (MeV)	SOURCE
PROTONS	0.1 TO 500	EARTH'S RADIATION BELTS AND SOLAR FLARES
ELECTRONS	UP TO 7	EARTH'S TRAPPED RADIATION BELTS
GALACTIC COSMIC RAYS	UP TO 20,000[†]	STEADY BACKGROUND IN DEEP SPACE (SLIGHTLY MODULATED BY SOLAR FLARE ACTIVITY)
NEUTRONS	0.001 TO 1000	TERRESTRIAL (FROM COSMIC RAY INTERACTIONS IN THE EARTH'S UPPER ATMOSPHERE)
ALPHA PARTICLES	UP TO 9	DECAY OF RADIOACTIVE IMPURITIES IN MATERIALS USED IN DEVICES

[†]A very small number of galactic cosmic rays have much higher energies.

Electrons and protons can also cause lattice damage, a bulk effect that moves atoms within the crystal away from their normal lattice positions. For many compound semiconductors bulk damage is more important than surface damage from ionization, in contrast to silicon MOS transistors, where surface damage dominates. The effects of electrons and protons on specific device technologies are discussed in Chapters 9 and 10. The remaining material in this chapter will discuss details of various radiation environments.

The distribution of particle energies is extremely important, determining how shielding within a spacecraft affects the final energy distribution of the particles after they pass through the outer skin and electronic enclosures of the spacecraft, as well as the effectiveness of the particles in producing damage effects, which depends on particle energy. Low-energy particles have limited range, and cannot penetrate the nominal 100-mil thickness of a typical aluminum enclosure used for electronics.

Electrons with energies up to 7 MeV, the range of electron energies within the earth's trapped belts, are more affected by shielding than trapped protons, which have a much higher maximum energy. For spacecraft in low-earth orbits, many of the particles that get through normal shielding are high-energy protons, diminishing the importance of effects from electrons (except for cases where there is less shielding). However, electrons are the dominant radiation source for high-altitude spacecraft, such as those operating in geosynchronous orbits.

Gamma rays are of minor importance in natural space environments, but they can be produced during interactions of other particles with shielding (*i.e.*, Bremsstrahlung radiation), or from on-board nuclear power sources. However, the main reason for considering radiation effects from gamma rays is radiation testing. Many radiation tests are done with laboratory gamma-ray sources because they are inexpensive and convenient. Radioactive decay of a cobalt isotope (^{60}Co) produces gamma rays with energies of 1.17 and 1.33 MeV. The gamma rays produce high-energy electrons when they interact with materials through the Compton scattering process. The electron energy distribution from that process peaks at about 500 keV; it is the interaction of the Compton electrons – actually created *within the material or nearby materials* by the Compton effect - that produces ionization effects in typical materials, not the gamma rays. The effect of gamma rays on semiconductor devices is not necessarily equivalent to the effects of particles in real space environments, which must be kept in mind in planning and interpreting radiation testing [1].

8.1.2 Particles producing transient effects

The last three particles in Table 8-1 – galactic cosmic rays for spacecraft, terrestrial neutrons, and alpha particles from radioactive impurities – are not present in sufficient numbers to cause the uniform type of displacement or ionization damage that results from the integrated effect of protons and electrons in space.

The main concern for galactic cosmic rays and terrestrial particles is spurious charge transients that are produced by the interaction of individual particles with a semiconductor. Those transients can disrupt normal operation, or even cause destruction for some device technologies if the spurious charge is high enough.

Protons produce short-duration transients in an analogous way, in addition to the integrated displacement and ionization damage that occurs from many proton interactions. The various effects that are produced by single-particle interactions are discussed in Chapter 13. However, we need to define the term that is usually used to compare ionization track effectiveness of various particles in order to interpret the space environment from the standpoint of transient effects on semiconductors. Linear energy transfer (LET), with units of MeV-cm^2/mg, is used for that purpose. It represents the energy loss of the particle, normalized to the density of the material.

8.1.2.1 Galactic cosmic rays

Galactic cosmic rays (GCR) are found everywhere in deep space. Their origin is not fully understood, but measurements show that they are comprised of hydrogen (the largest component) along with other atomic nuclei. There is a distribution of particle energies, up to 10^{20} eV, an extremely high energy. For energies $<10^6$ GeV the differential flux has been found to obey a power-law dependence [2],

$$J(E) = \text{const} \times (1+E)^{-2.75} \qquad (8\text{-}1)$$

where J(E) is the integral intensity in $(\text{cm}^2\text{-s-ster-eV})^{-1}$, and E is the cosmic ray energy. Above 10^6 GeV the slope is much steeper, but there are very few particles with energies above that value. Equation 8-1 shows that the intensity drops sharply with energy. The number of cosmic rays with energies above 1,000 GeV is high, $\sim 10^5$ $\text{cm}^{-2}\text{-yr}^{-1}$. Such energetic particles have extremely long range, and can pass through the normal amounts of shielding that are present in spacecraft.

The power law relationship for cosmic ray flux can be explained by a theory that assumes that the cosmic rays are gradually accelerated by various magnetic sources within the universe, including shock waves from supernova explosions [2,3]. When cosmic rays enter the galaxy, they are trapped for an average time period of 10^6 years, a time interval that is much shorter than the mean time for collisions with interstellar hydrogen. Some particles escape at the galactic boundary, but the long lifetime allows them to be accelerated to very high energies during the time that they are trapped. The mechanisms for acceleration are still being debated, particularly for particles with energies above 10^6 GeV [4].

From the standpoint of radiation effects in electronics we need to be concerned about isotopic abundance as well as the energies of cosmic rays. There is a wide distribution of ion types in the GCR particle spectrum, roughly mirroring the distribution of elements in the earth's crust (the galactic particles are formed through spallation reactions, and the ionic abundance is determined by the stability of the various nuclei [2,3]). Most of the particles have atomic numbers below $Z = 26$ (iron). Many studies of GCR spectra in space have used particle detectors with three energy channels: one detecting alpha particles, one detecting particles with intermediate atomic number (dominated by C, N and O), and a higher Z channel that is dominated by iron. There have also been studies using better resolution that provide more complete information about the composition of galactic cosmic rays [5]. Although cosmic rays with low energies can be shielded, there are large numbers of ions with extremely high energies that are only slightly affected by shielding.

Galactic cosmic rays are completely ionized. When a cosmic ray interacts with a semiconductor or insulator, an intense ionization wake is produced along the particle track. The initial track diameter is about 0.1 μm, but it spreads laterally due to ambipolar diffusion, a consequence of the very high density of holes and electrons that are generated within the track [1]. The track diameter increases to about 1 μm about 1 ns after the particle strike.

The deposited charge per unit path length, an important metric for evaluating the effects of the interaction, is proportional to the square of the atomic number, Z, of the incident particle.

For highly energetic particles, near the so-called minimum ionization region, energy loss for all types of nuclear particles can be normalized to Z^2. The ratio of this normalized energy loss to the energy/nucleon ratio is shown in Fig. 8-1 for helium and iron. The E/m ratio of particles in the minimum ionization region is approximately 1000 MeV/nucleon, a very high ratio, but the ionization loss value is low.

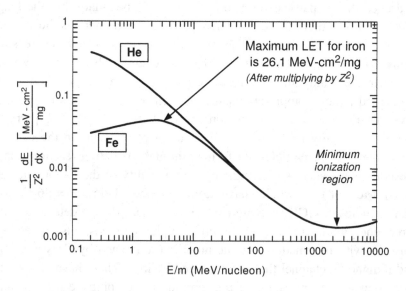

Fig. 8-1. Normalized energy loss for alpha particles and iron. The data are normalized to Z^2, resulting in a nearly universal curve for E/m > 30 MeV/nucleon.

The energy loss rises steeply for particles with energies below the minimum ionization region, and the Z^2 normalization is no longer effective when the E/m ratio is <30. For each ion type there is a distribution of energies. The peak in the distribution of GCR energies occurs at approximately 200 MeV/nucleon, where the LET is slightly higher compared to the minimum ionization value.

In nearly all cases, the LET of particles in the minimum ionization region is below the threshold LET level that is required to cause upsets in working devices. For example, if we remove the $1/Z^2$ normalization factor, the energy loss for iron (Z = 26) in the minimum ionization region – expressed as MeV-cm^2/mg – is almost exactly 1 MeV-cm^2/mg. However, cosmic rays with lower energy – or particles that lose energy as they travel through extended thicknesses of material within a spacecraft – will have higher LET values compared to the minimum ionization value.

For example, the maximum LET for iron (Z = 26) is 26 MeV-cm^2/mg. That specific LET value is often used as an effective cutoff value for single-particle effects in space because there are so few ions with Z above 26. However, it is important to remember that other atomic species with higher LET are also present in the GCR environment. The effects of those ions can be significant even though their abundance is lower than that of iron, particular for long-duration missions.

8.1.2.2 Solar flares

Solar flares, or more properly coronal mass ejections from the sun, are roughly correlated with sunspot activity. The flares produce energetic protons as well as heavy ions that travel along the sun's magnetic field lines. The majority of solar flares are not earth-directed, and most solar flares have no effect on earth-orbiting spacecraft. However, deep space missions can be hit by flares that do not affect the earth, and are not detected by spacecraft that monitor solar flare activity in the earth's vicinity.

The limited information about solar flare activity for those missions increases the risk of failure, and must be taken into account in operational requirements if a "safehold" condition occurs because of failures or disruptions in the electronics.

The path of some solar flares is directed towards the earth. Intense flares can degrade solar cells, affecting the power system, as well as producing damage and single-event upset in the internal electronics. The flares can persist as long as 24 hours, and may interrupt normal operations.

The distribution of atomic numbers of the particles in a solar flare is similar to that of galactic cosmic rays [7]. However, solar particles have lower energies compared to GCR particles, and are more affected by shielding. The energy distribution is not the same for all flares. Some flares have much higher energies, increasing the relative number of particles that penetrate shielding compared to "soft" flares. The integral proton fluence is often used as a metric for solar flare intensity because protons are easily measured, and occur in larger numbers than the heavy solar particles.

8.1.2.3 Terrestrial radiation

Terrestrial particles include alpha particles, produced by the radioactive decay of trace impurities in packaging materials, metallization, and refractory metals; and neutrons. Neutrons are produced when galactic cosmic rays interact with nitrogen in the upper atmosphere [6]. Because they have no charge, the interaction probability is small, with a long mean-free path through the atmosphere. A number of them survive, producing a small but significant fluence of neutrons at the earth's surface.

These atmospheric neutrons can interact with semiconductors in normal terrestrial applications, producing upsets in a manner that roughly parallels the effects of protons in space, although the upset rate is much lower. The neutron fluence varies somewhat with location at the earth's surface. It also increases with altitude; the neutron fluence is about 200 times higher in an aircraft operating at 35,000 feet compared to the fluence at sea level.

Neutron interactions, along with interactions from alpha particles, are of great concern for the semiconductor industry because the sensitivity of electronic devices to such interactions increases markedly for devices with small feature size and fast switching time. The sensitivity of advanced devices has reached the point where most commercial manufacturers consider these effects in device design, and include maximum soft-error rates in device specifications [7].

8.2 Radiation Environments Near the Earth

The earth's magnetic dipole can trap high-energy charged particles. The particles spiral along the magnetic field lines, and are reflected when they reach the region near the poles. Figure 8-2 shows a highly simplified diagram of the location of the Earth's radiation belts. The distance between these regions is usually described in terms of the earth's radius, R_e (6380 km). Unlike the simple diagram in the figure, the radiation belts are asymmetric, extending about 12 R_e in the direction towards the sun, and to 40 R_e away from the sun.

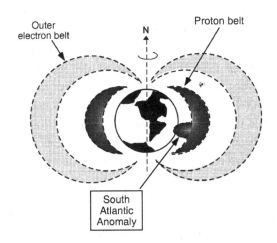

Fig. 8-2. Simple diagram of the Earth's trapped radiation belts.

There are three radiation belts: an inner electron belt, a proton belt, and an outer electron belt. The position of the belts is often discussed based on the altitude at the equator, but this is somewhat misleading because the belts curve at high latitude.

The geometrical details of the trapped belts are far more complicated than indicated by this simple diagram (see Ref. [8] for more details). The proton and electron belt contours are not constant with altitude, which means that spacecraft that orbit at a fixed altitude actually pass through both the proton and electron belts, unless they are in a low-inclination orbit. Consequently we have to be concerned about both protons and electrons for spacecraft in low-earth orbits.

The inner electron belt, not shown in the figure, extends from the upper atmosphere to about 7 km (at the equator). Electrons within the inner belt have lower energy than electrons in the outer belt, reducing its importance from the standpoint of electronics in spacecraft. We will not discuss it further because effects from the proton belt are usually more important for spacecraft in low earth orbits. The proton belt extends from approximately 1 R_e to 3.8 R_e (6,400 to 24,000 km) at the equator. The total dose from protons increases rapidly with increasing altitude. For that reason, the majority of low-earth orbit spacecraft operate at altitudes below 1,500 km.

There is one more important feature, the South Atlantic Anomaly (SAA). This region, where the trapped proton belts extend to much lower altitudes, occurs roughly over Brazil. It is caused by two factors. First, the Earth's rotational axis and the axis of the magnetic dipole that is responsible for the trapped radiation belts are slightly misaligned, differing by about 11°. Second, there are inhomogeneous regions in the magnetic material in the earth's crust that also contribute to the "dip" in the field lines. Those factors distort the proton belt, as indicated in a very approximate way in Fig. 8-2. The distorted region extends to altitudes as low as 200 km.

Spacecraft that orbit the earth at altitudes below the edge of the trapped proton belt will be temporarily exposed to a high flux of protons during the short time that they traverse the SAA region. The SAA is highly localized for altitudes of a few hundred kilometers, but extends over a much broader region at higher altitudes. Protons from the SAA are usually the dominant radiation problem for spacecraft in low earth orbits For example, trapped protons caused malfunctions in some of the circuits in the Hubble space telescope when it passed through the South Atlantic Anomaly. The problem was fixed by installing upgraded electronics on one of the Hubble servicing missions.

Electrons from the outer electron belt are the dominant radiation source for high-altitude missions, including geosynchronous orbits (~36,000 km). Even though there are no proton belts in that region, protons from solar flares also have to taken into account when the environmental requirements are established for spacecraft at that altitude.

8.3 Energy Distributions in the Earth's Trapped Belts

For low-earth orbits, protons are the most important constituent of the Earth's trapped radiation belts. The energy distribution extends to very high energies, >500 MeV. The differential energy distribution – dN/dE vs. energy – is shown in Fig. 8-3 for two orbits, a 705 km, high-inclination orbit, which has been used by several spacecraft that monitor weather and other features of the earth; and an orbit at 1334 km (with lower inclination), used for several missions that monitor features of the ocean. The 1334-km orbit is high enough to encounter the lower edge of the proton belt during part of the orbit, but both orbits are heavily influenced by the SAA. The intensity of protons in the 1334 km orbit is about 20 times larger than that of the lower orbit.

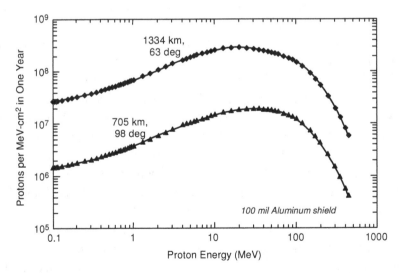

Fig. 8-3. Differential proton spectra behind a spherical 100-mil aluminum shield for two low-earth orbits.

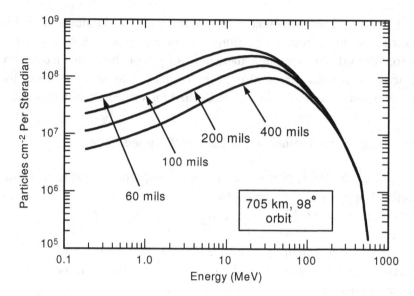

Fig. 8-4. Proton energy distribution for various spherical shields for a 705 km high-inclination earth orbits. The shape is quite different for very thin shields, with a relatively higher fluence at low energy.

There is a wide distribution of proton energies for both orbits. Shielding affects the relative number of protons at lower energies, but incremental amounts of shielding are less effective once the initial low energy protons are removed. The reason for this is that shielding reduces the energy of some protons with higher energy, and those protons replace low energy protons that are removed by shielding. The net effect is that the shape of the distribution does not change, but the overall number of protons decreases. Figure 8-4 shows the effect of different spherical shield thicknesses on the proton energy distribution for the 705 km orbit in Fig. 8-3. The peak energy increases as more shielding is added; the peak energy distribution occurs at about 15 MeV for a 60-mil shield, but it increases to about 40 MeV for a 400-mil shield.

The energy distribution turns out to be even more important than suggested by these figures, because protons with lower energy are much more effective in producing displacement and ionization effects. The mean energy and the total proton fluence per year are useful benchmarks when comparing different environments, but they have to be corrected for the energy dependence of total dose and displacement mechanisms in semiconductors in order to make meaningful comparisons for their effects on spacecraft electronics.

There is a similar energy distribution for electrons. However, as discussed in Section 8.1, electrons in the earth's trapped belts have lower energies than protons, and are easier to shield. The energy dependence of radiation damage effectiveness is less important for electrons, and we will only consider the net effect of shielding for the electron environment, not the energy distribution.

8.4 Radiation Environment in a Geosynchronous Orbit

Some earth-orbiting spacecraft are placed in a geosynchronous orbit, at an altitude of 35,786 km above the earth. At that altitude a spacecraft with 0° inclination (*i.e.*, within the equatorial plane) will appear stationary at a fixed position at the equator. If the inclination is not zero, the path of the spacecraft will appear to be a "figure 8" over the earth's surface, a geostationary rather than a geosynchronous orbit. Geostationary orbits are subject to other forces that cause the spacecraft inclination to drift, requiring occasional corrections ("station keeping") that are not required for geosynchronous orbits.

The 35,786 km orbit, which corresponds to about 6.6 earth radii, is within the outer electron zone [8]. Electrons within that zone have a maximum energy below 7 MeV. Shielding is relatively effective in reducing the total dose for this type of orbit. Figure 8-5 shows the cumulative total dose for a geosynchronous spacecraft that is required to operate for five years.

Although the total dose is extremely high for very thin shielding, it is reduced to about 50 krad(Si) behind a 100-mil spherical shield. Adding additional shielding can reduce the total dose for the entire mission length to about 10 krad(Si). However, when the shield thickness is >300 mils, the gamma rays from Brehmsstrahlung place a lower limit on the effectiveness of shielding, as shown in the figure (gamma rays are only slightly affected by "normal" shielding thicknesses).

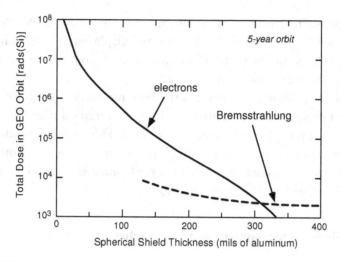

Fig. 8-5. Total dose vs. spherical aluminum shield thickness for a 5-year geosynchronous orbit.

8.5 Protons from Solar Flares

The environment in deep space, on average, has a very low level of high-energy protons. Spacecraft that travel beyond the earth's radiation belts will be exposed to very low daily radiation levels, except during a solar flare. The proton environment in deep space is dominated by solar flare activity, which is highly variable, and can only be treated statistically because of the random nature of solar events.

8.5.1 Solar activity

The sun is a very complex system that produces solar flares (more accurately, coronal mass ejections) that occur at random times. Despite the random nature of the events, they are correlated with an overall eleven-year cycle[*], which consists of four relatively quiet years followed by seven active years. Each eleven-year cycle produces hundreds of

---

[*] The solar cycle is really 22 years, not eleven. The sun's geomagnetic poles change direction between successive 11-year cycles. Nevertheless, periods of peak sunspot activity can be correlated with an 11-year cycle.

flares, most of which have relatively low intensity. The flares are roughly correlated with sunspot activity, but not all flares occur in conjunction with a sunspot.

The charged particles from a solar flare travel in a spiral path along the sun's magnetic field lines. Most flares are not earth-directed, but the small percent that reach the earth (or the position of a spacecraft that is located beyond the earth) can disrupt spacecraft electronics and produce permanent damage. Intense flares can persist for several days, interfering with normal spacecraft operation for an extensive time period unless the spacecraft is designed to withstand such events.

Measurements of solar flare intensity have been done for more than 40 years. Although the energy distribution varies for different flares, the fluence of protons with E > 10 MeV is often used as a relative measure of flare intensity. Proton fluence data for individual flares has been fitted to a log-normal distribution by Feynman and Gabriel [9]. Their model predicts that very high intensity flares – that is, flares with a cumulative 10-MeV fluence $>10^{10}$ p/cm^2 – are relatively rare. The integrated fluence of a "typical" intense flare is in the range of 2–3 x 10^9 p/cm^2. This level is about an order of magnitude lower than the fluence from a worst-case flare. It is possible to develop predictions for specific missions from such models, assuming certain confidence limits, and that approach is often used to develop total dose requirements for space missions.

8.5.2 Specifying proton flare environments

In discussing solar flares, we have to carefully distinguish between the maximum fluence from a single very intense flare that will persist for about 24 hours, and the integrated effect of large numbers of flares with lower intensity over the entire mission length. For total dose and displacement damage, the integrated result is usually more important than the effect of a single intense flare. However, the peak intensity of a single flare is usually more important for transient effects (see Chapter 13).

It is relatively easy to measure proton intensity at specific cutoff energies, using specialized instrumentation with different shielding configurations. The intensity of a solar flare is often characterized in this way, even though we also have to be concerned about heavy ions produced by the flare as well as protons. Statistical models have been developed for solar flares using a 10-MeV cutoff energy as a metric [9,10]. The distribution of flare intensities fits a log-normal distribution, with a peak fluence of about 3×10^{10} p/cm^2 for the most intense flare. This statistical approach is useful for predictions of intensity. The dashed lines in Fig. 8-6 (after Xapsos, *et al.*) show the predicted total dose vs. spherical shell shielding thickness for flares of different intensity, using the log-normal distribution to determine what fraction of the total number of flares will have an intensity below the value indicated by the dashed lines [10]. The model also assumes that the proton energy distribution is identical for all flares.

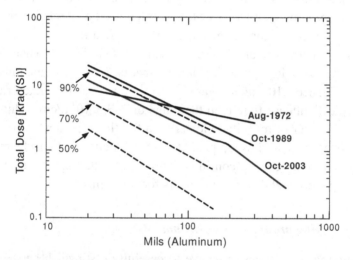

Fig. 8-6. Cumulative total dose from three intense solar flares, along with predicted values from statistical solar flare proton models [10]. The August 1972 flare has much higher energy compared to most flares, reducing the effectiveness of shielding. New experimental results for the October, 2003 flare have been added to the figure. © 2000 IEEE. Reprinted with permission.

Many solar flares, including the intense flare in October, 1989, have closely followed that dependence. However, some flares, including the August, 1972 flare, have far more high energy protons, making shielding less effective. Measured results for the October, 2003 flare, which roughly corresponds to an "85%" flare, are also included in Fig. 8-6.

Xapsos, *et al.*, developed a design model for the cumulative effect of solar flares over longer time intervals, based on statistical modeling [11]. Figure 8-7 plots the integrated total dose from the cumulative effect of all flares that occur in a seven-year cycle vs. shielding thickness for various confidence limits using their model. Measured results for cycle 22 (1986–1997) are also included. Several intense flares occurred during that particular cycle, and the measured results closely correspond to the 70% confidence limit of the model. A 90% or higher confidence limit is often used to develop requirements for space systems.

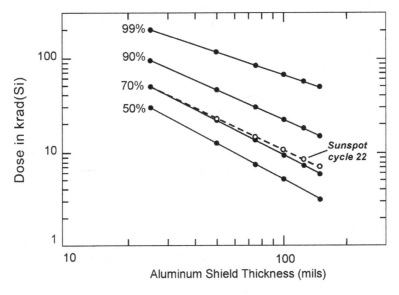

Fig. 8-7. Total dose from the cumulative effect of flares during the seven year active period of a solar cycle for various confidence limits [10]. © 2000 IEEE. Reprinted with permission.

The total dose profile of a space mission depends on the launch date as well as the total operating time. A time profile for high-energy protons that can be used for an eleven-year mission (the model covers one complete solar cycle) is shown in Fig. 8-8, starting during the quiet time, and assuming a 90% confidence limit [11]. The time profile is shown for cumulative fluence within various energy ranges for all of the flares that are expected during an extended mission. The cumulative fluence for protons with E > 4.9 MeV is about 2×10^{11} p/cm^2. Note the abrupt increase in cumulative fluence between the 4th and 5th years, due to the transition between the quiet and active period of the sun that is assumed in this model.

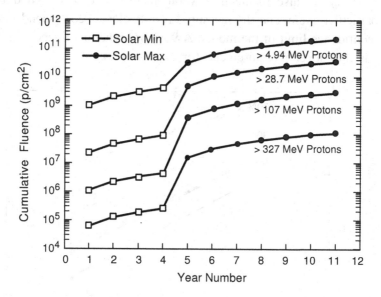

Fig. 8-8. Cumulative proton fluence for different energy ranges vs. time for an eleven-year mission that starts during the beginning of an eleven-year solar cycle [11]. A confidence level of 90% was assumed. © 2004 IEEE. Reprinted with permission.

The fluences in Fig. 8-8 are based on a statistical model that is somewhat conservative, but does not represent worst-case conditions. The fluence profile during a specific solar cycle may be quite different, particularly if a very high intensity solar flare occurs. The fluence from a single high-intensity flare could be up to 20% of the cumulative fluence expected from this model over the entire eleven-year cycle.

Although statistical models are useful, they can only predict solar flare intensities with relatively large statistical error bars. For example, a relatively intense flare occurred in mid-December, 2006, during the middle of the "quiet" period of solar cycle 23.

8.6 Galactic Cosmic Rays

As discussed in 8.1.2, most galactic cosmic rays have extremely high energies, and are only slightly affected by shielding. For compound semiconductors, the most important effect from these particles is ionization, which produces a dense charge track along the path of the particle. As discussed earlier, the parameter used to describe the charge density along the track is linear energy transfer (LET), with units of MeV-cm^2/mg. It is essentially the amount of energy lost by the particle along the path as it travels within a specific material.

LET can be described differently – charge induced per unit path length, in units of pC/μm – which is more intuitive when describing device behavior. We will use LET* when expressing LET in units of pC/μm.

Although LET* is more straightforward when we consider charge deposition in electronic devices, the density of the material must be included in making a conversion between units. As a result, conversions between LET and LET* are not the same for different materials. For silicon, an LET of 10 MeV-cm^2/mg corresponds to 0.104 pC//μm (~1/100), whereas for GaAs the equivalence is 0.0177 pC//μm (~1/60) because of the higher density of GaAs.

Figure 8-9 shows the galactic cosmic ray spectrum in deep space, along with the spectrum for two earth orbits. In deep space the number of particles decreases by about four orders of magnitude as the LET increases from 20 to 30 MeV-cm^2/mg. This is sometimes referred to as the "iron threshold" because only particles with Z > 26 can produce LET values above 26 MeV-cm^2/mg, and they are relatively rare.

For earth orbits the earth's magnetic field provides shielding from highly energetic particles, except near the poles. The 98° orbit is directly exposed to cosmic rays during the time that the orbit is near the poles, but magnetic shielding reduces the number of particles during much of the orbit. The net effect is to reduce the number of particles by about a factor of four compared to deep space, with little effect on the overall LET distribution. Geomagnetic shielding is far more effective for the low inclination 28° orbit, effectively eliminating particles with LET values above 15 MeV-cm^2/mg from the distribution.

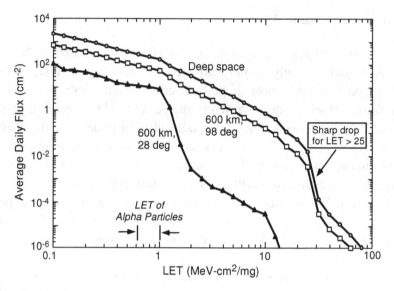

Fig. 8-9. Galactic cosmic ray energy distribution in deep space compared to two earth orbits with different inclination.

The essential point of Fig. 8-9 is that we have to be concerned with the entire spectrum of galactic cosmic ray particles for earth-orbiting spacecraft unless the orbit has an inclinsation below approximately 35°.

Galactic cosmic rays have extremely high energies, and are only slightly affected by shielding. However the solar wind has a long-range effect on the GCR particle distribution which causes an *inverse* correlation between solar activity and the GCR intensity [13]. During periods of low solar activity the GCR intensity is four to five times greater than during periods of high solar activity. This solar modulation effect is important for nearly all space missions, including those extending to the outer planets.

Except for the slowly varying solar modulation effect, the flux of GCR particles is remarkably constant. However, the effect on a particular device depends on the probability that an ion with an LET value that is high enough to cause the part to respond actually strikes the sensitive area. Despite the near constant environment, the small number of particles that actually strike a typical device means that upset from GCR can only be determined within statistical limits.

8.7 Heavy Particles in Solar Flares

Heavy ions from solar flares have essentially the same effect on semiconductor devices as GCR particles, although ions from solar flares have lower energy. For an intense flare, the flux can be large enough to increase the number of upsets by as much as four orders of magnitude for several hours, compared to the "background" upset rate from GCR. Space systems must be designed to withstand the high peak rate of upsets (or other SEU-related effects) from an intense flare [14,15].

As discussed earlier, there is a wide variation in the intensity of the radiation from solar flares that applies to the flux of heavy particles as well as to protons. The probability of an extreme solar flare is relatively low, but the usual practice is to design electronics so that they can withstand the high intensity of a very severe flare.

Although a great deal of information is available about the intensity and energy distribution of protons, less is known about heavy particles in solar flares. Tylka, *et al.*, analyzed solar flare data from the IMP-8 satellite, taken between 1973 and 1996 [14]. Three basic channels of data were available: one detecting He; one detecting C, N and O; and a third that detected Fe. Each channel was replicated, adding extra shielding to one channel of each type in order to distinguish between low and high energy particles. Figure 8-10(a) shows the low-energy results for the middle channel, which is sensitive to C, N and O. Note the large difference for this channel for the average flux between the worst day and the GCR fluence.

Results for this same channel at high energies (with less shielding) are shown in Fig. 8-10(b). With increased shielding a smaller increase in intensity occurs relative to the steady GCR background level compared to the case where less shielding is used. Only about 1% of the solar flares have sufficient flux to exceed the background level.

For the low-energy channel in Fig. 8-10(a) the average flux in a "worst day" is about four orders of magnitude greater than the daily flux from GCR. In comparison, the flux in a "worst day" in the active channel is only about two orders of magnitude above the average daily flux. There are similar differences between the average daily GCR flux and the flux from a worst day for iron [16]. These results show the importance of shielding for the solar flare environment.

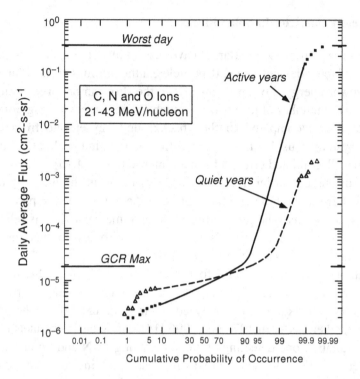

Fig. 8-10(a). Probability distributions for a low-energy measurement channel that includes carbon, nitrogen and oxygen, based on experimental observations over two solar cycles (after Tylka, et al. [14]). © 1997 IEEE. Reprinted with permission.

Solar flare energy and LET distributions have been included in models of space environments, such as CRÈME96 [17]. A comparison of upset rates from the GCR background and the upset rate for a very large flare (in this case, the October, 1989 flare) is shown in Fig. 8-11, using the CRÈME96 model. The vertical axis shows the relative upset rate, normalized to the upset rate (from GCR) of a device with an LET threshold of 1 MeV-cm^2/mg with a cross section of 1 cm^2. A step function is assumed for the cross section at the LET threshold.

Device upset rates can be estimated by taking a "cut" on the horizontal axis to determine the threshold LET, and multiplying the vertical axis by the measured (or calculated) saturation cross section for the specific device, which is usually much smaller than 1 cm^2.

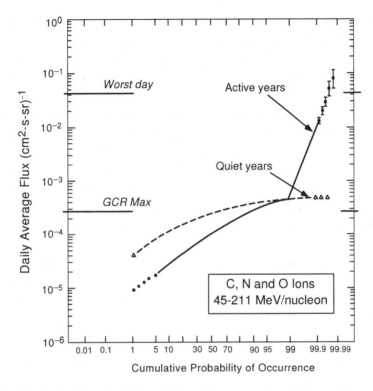

Fig. 8-10(b). Probability distributions for a measurement channel that includes carbon, nitrogen and oxygen, based on experimental observations over two solar cycles (after Tylka, *et al.* [14]). © 1997 IEEE. Reprinted with permission. These data are for a detector with more shielding than the results in Fig. 8-10(a), and provide data for this same channel at higher particle energies.

The relative upset rate during a very intense flare in a spacecraft with an effective shielding thickness of 60 mils (Al) is shown in the top curve. For a very sensitive device (LET threshold = 1 MeV-cm^2/mg), the upset rate during the approximate 24-hour interval of the flare is more than four orders of magnitude higher than the rate due to the GCR background. With a relatively thin shield, the relative increase in upset rate is nearly the same for devices with high threshold LET values.

First, consider the curves for GCR, which differ by about a factor of four because of solar modulation. The relative upset rate decreases for devices with higher threshold LET values because the number of particles decreases (see Fig. 8-9). The GCR curve for solar maximum (dashed line) is the environment that is expected during periods of high solar activity, and we will use that as a reference for discussing the enhanced environment during solar flares.

If we compare that result to the 250 mil shield case, the peak upset rate is only about 1000 times higher from the flare for a device with LET threshold = 1 MeV-cm^2/mg. With that shielding assumption, the relative increase in upset rate falls to about 250 for a device with LET threshold = 35 MeV-cm^2/mg. These differences are due to the lower energy of heavy particles from solar flares which are more affected by shielding (the GCR background is essentially unaffected by shielding).

The example in Fig. 8-11 is for a very intense flare. Most flares have far lower intensities, but even a moderate flare can increase the SEU rate by two orders of magnitude compared to the GCR background rate, as shown by the statistical probability curves in Figs. 8-10(a) and 8-10(b).

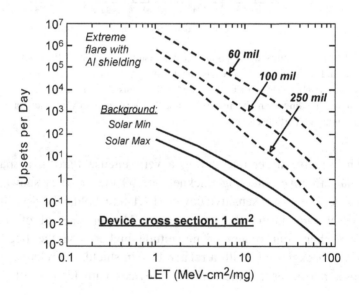

Fig. 8-11. Comparison of upset rates from an intense solar flare with the upset rate from GCR background.

8.8 Terrestrial Environments

Relatively low levels of radiation occur in terrestrial environments. The levels are low enough to essentially eliminate concern about long-term damage effects. However, transient effects can be important, particularly for high-speed devices that require very small amounts of charge for switching. For mainstream silicon microelectronics, the combination of increased sensitivity to soft errors along with very high integration density has elevated the importance of these effects to the point where they are the largest single factor for reliability. Compound semiconductors usually have lower levels of integration, reducing the overall cross section for soft errors, but their high speed makes them susceptible to soft error mechanisms at low LET. We will discuss the details of these effects in Chapter 13, but the environmental features are discussed in this chapter.

8.8.1 Alpha particles

Various radioactive materials are present in semiconductors, including actinide elements that emit alpha particles through normal radioactive decay (the nuclei are unstable, with long half lives). The impurities are present in very low concentrations in metallization, contact metals, packages, and even within the semiconductor material. For example, Fig. 8-12 shows the energy of various alpha particles emitted by uranium-238. The different alpha energies are the result of different decay channels, not shielding. The initial LET of the alpha particle ranges from 0.5 to 0.7 MeV-cm^2/mg. The LET increases as the particle loses energy, and can have a maximum LET of 1.6 MeV-cm^2/mg, depending on how far the alpha particle travels through the material before it reaches a sensitive location. That LET value is high enough to cause upsets in some types of high-speed devices.

To keep the upset rate below critical levels, most materials used to fabricate semiconductor devices are required to meet specifications that limit alpha particle emission. In addition to the material specifications, the maximum soft-error rate, in errors per bit day, is often specified in the data sheet for the part, even for parts that are used in commercial applications.

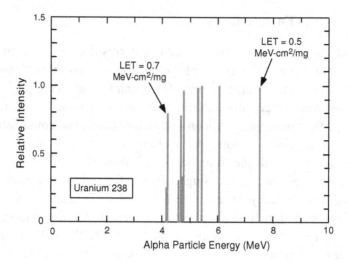

Fig. 8-12. Alpha particle spectrum from uranium-238, a common radioactive impurity that is present in low concentrations in some materials used within semiconductors.

In addition to spontaneous decay, neutrons and protons can also initiate nuclear reactions. This leads to the possibility of higher upset rates in space applications, particularly during an intense solar flare.

Upsets from alpha particles are usually less important for compound semiconductors compared to silicon semiconductors because of the lower integration density. However, the high speed and low switching margins of compound semiconductors can make them even more susceptible to these mechanisms.

8.8.2 Atmospheric neutrons

As discussed at the beginning of this chapter, neutrons are generated in the upper atmosphere when cosmic ray protons interact with nitrogen. Some of the neutrons reach the ground, creating a low neutron radiation level that can affect high-speed electronic devices. The neutron flux increases inversely with energy; it is about three orders of magnitude higher at 1 MeV compared to the flux at 100 MeV. This causes the upset rate to increase rapidly as devices are scaled to smaller feature size, because the critical charge to upset them generally decreases. For high-speed devices upset from atmospheric neutrons can be even more important than upsets from alpha particles in commercial microelectronics.

The energy distribution of atmospheric neutrons at ground level is shown in Fig. 8-13 [7]. It is interesting to contrast this with the proton energy distribution for space environments, shown previously in Fig. 8-4. Most low-energy protons are effectively removed by shielding from thin enclosures, which causes the proton spectrum in space to decrease at low energies. That is not the case for terrestrial neutrons; the relative number of neutrons increases rapidly at low energy. Consequently the sensitivity of devices to neutron upset will increase significantly if the design of the devices changes in a manner that allows them to upset at lower neutron energies. Such changes could occur from a decrease in critical charge, or by the addition of materials in device structure or packaging that have significant reaction cross sections at low neutron energy.

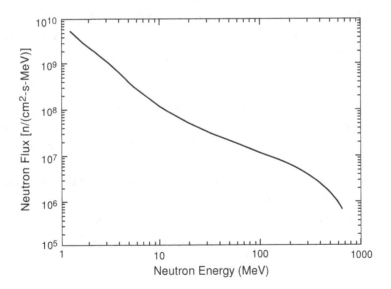

Fig. 8-13. Energy spectrum of atmospheric neutrons at ground level [7]. The spectrum varies slightly with latitude, but this spectrum is representative of the spectrum at sea level. © 2005 IEEE. Reprinted with permission.

Neutrons are found in some space environments, particularly in space probes to the outer planets that rely on radioactive power sources. However, in nearly all cases upsets from the natural space environment will be the dominant contribution for space systems. The International Space Station is an exception, because the structure is so massive compared to a typical spacecraft.

There is another point of view. Concern about the effects of atmospheric neutrons is potentially beneficial for devices that are used in space applications. Devices that are designed to keep neutron upset within bounds in terrestrial applications have some rudimentary level of radiation hardening. This will reduce their sensitivity to upset from protons in space environments, providing a floor on susceptibility.

8.8.3 Nuclear reactors

Although it is a special application, very high radiation levels can be encountered in monitoring and control systems for nuclear reactors. In most cases active electronics are not used near the reactor core, but are installed outside shielding in order to reduce the radiation level in active electronics to acceptable levels. The specific requirements vary widely, depending on the reactor design, but they are usually much higher than the requirements for space systems.

Gamma radiation is present in reactors, and total dose levels above 1 Mrad(Si) are often specified for reactor electronics. The electronics must also be capable of withstanding high levels of fission neutrons (mean energy approximately 1 MeV). In addition to the radiation requirement, reactor electronics frequently must operate at elevated temperature.

Compound semiconductors are excellent choices for this demanding environment for several reasons. SiC and GaN can operate at much higher temperature than conventional semiconductors because of their wide bandgaps. As discussed in the next chapter, they are also very resistant to gamma radiation.

8.9 Summary

This chapter has presented an overview of various radiation environments. Most of the discussion centered around two types of space missions: earth-orbiting missions, where trapped radiation in the Van Allen belts is the main contributor to the radiation environment encountered by spacecraft; and deep space missions, where galactic cosmic rays are the primary concern. Both types of missions are affected by solar flares, which occur at random times with a wide variation in intensity.

Models are available for earth orbits that can calculate the total dose, particle fluence and energy distribution for electrons and protons, starting with the orbital parameters (including shielding) of a particular spacecraft. They are usually used to establish environmental requirements for a particular mission. The models also contain statistical models for solar flares.

For deep space, total dose typically depends on solar flares, which occur at random times with various intensities. Statistical models can be used to establish appropriate values for solar flares, which produce a large increase in the number of heavy particles that cause transient effects in addition to total dose and displacement damage from protons.

A brief treatment of terrestrial radiation environments was also included, even though this is of less concern for most compound semiconductors compared to silicon technology. It is important to realize the potential importance of terrestrial radiation problems for high speed devices because they affect the overall reliability of devices that are used in conventional applications. The next chapter will discuss the way that particles in space interact with materials and devices.

References

1. J. R. Srour and J. M. McGarrity, "Radiation Effects on Microelectronics in Space", Proc. IEEE, **76**(11), pp. 1443–1469 (1988).
2. P. Morrison, S. Olbert and B. Rossi, "The Origin of Cosmic Rays", Phys. Rev., **94**(2), pp. 440–453 (1954).
3. J. W. Cronin, "Cosmic Rays: The Most Energetic Particles in the Universe", Rev. Mod. Phys., **71**(2), pp. S165–S175 (1999).
4. V. L. Ginsburg and V. S. Pluskin, "On the Origin of Cosmic Rays: Some Problems in High-Energy Astrophysics", Rev. Mod. Phys., **48**(2), pp. 161–189 (1976).

5. D. L. Chenette and W. F. Dietrich, "The Solar Flare Heavy Ion Environment for Single-Event Upsets: A Summary of Observations Over the Last Solar Cycle, 1973-1983", IEEE Trans. Nucl. Sci., **31**(6), pp. 1217–1222 (1984).
6. J. F. Ziegler and W. A. Lanford, "Effect of Cosmic Rays on Computer Memories", Science, **206**, pp. 706–788 (1979).
7. R. C. Baumann, "Radiation-Induced Soft Errors in Advanced Semiconductor Technologies", IEEE Trans. on Device and Materials Reliability, **5**(3), pp. 305–316 (2005).
8. E. G. Stassinopoulos and J. P. Raymond, "The Space Radiation Environment for Electronics", Proc. IEEE, **76**(11), pp. 1423–1442 (1988).
9. J. Feynman and S. Gabriel, "High-Energy Charged Particles in Space at One Astronomical Unit", IEEE Trans. Nucl. Sci., **43**(2), pp. 344–352 (1996).
10. M. A. Xapsos, *et al.*, "Characterizing Solar Proton Energy Spectra for Radiation Effects Applications", IEEE Trans. Nucl. Sci., **47**(6), pp. 2218–2223 (2000).
11. M. A. Xapsos, *et al.*, "Model for Solar Proton Risk Assessment", IEEE Trans. Nucl. Sci., **51**(6), pp. 3394–3398 (2004).
12. L. W. Townsend, E. N. Zapp, D. L. Stephens, Jr., and J. L. Hoff, "Carrington Flare of 1859 as a Prototypical Worst-Case Solar Energetic Particle Event", IEEE Trans. Nucl. Sci., **60**(6), pp. 2307–2309 (2003).
13. I. Sabbah and M. Rybansky, "Galactic Cosmic Ray Modulation During the Last Five Solar Cycles", J. Geophys. Res., **111**. A01105, doi:10.1029/2005JA011044 (2006).
14. A. J. Tylka, W. F. Dietrich and P. R. Boberg, "Probability Distributions of High-Energy Solar-Heavy_Ion Fluxes from IMP-8: 1973–1996", IEEE Trans. Nucl. Sci., **44**(6), pp. 2140-2149 (1997).
15. P. P. Majewski, E. Normand and D. L. Oberg, "A New Solar Flare Heavy Ion Model and Its Implementation Through MACREE, An Improved Modeling Tool to Calculate Single Event Rates in Space", IEEE Trans. Nucl. Sci., **42**(6), pp. 2043–2050 (1995).
16. J. H. Adams, Jr., "The Variability of Single Event Upset Rates in the Natural Space Environment", IEEE Trans. Nucl. Sci., **30**(6), pp. 4475–4480 (1983).
17. A. J. Tylka, *et al.*, "CREME96: A Revision of the Cosmic Ray Effects on Microelectronics Code", IEEE Trans. Nucl. Sci., **44**(6), pp. 2150–2160 (1997).

Chapter 9

Interactions of Radiation with Semiconductors

9.1 Fundamental Interactions

As discussed in the previous chapter, the energetic particles in space that we are concerned with for permanent damage in electronics include protons, electrons, and heavy charged particles. Neutrons may also be included when atomic power sources are present on spacecraft, or if we consider effects in nuclear reactors.

Neutrons produced in the upper atmosphere can also be important in terrestrial applications for highly scaled integrated circuits, but do not occur in sufficient numbers to cause permanent damage, only transient effects.

Several different physical processes are involved when these particles interact with semiconductors, which are quite complicated to treat in detail. We can simplify our discussion by emphasizing the final effects of the particles on semiconductor devices, provided only a limited treatment of fundamental interactions and processes. The material in this chapter will mainly address the effects that cause integrated damage. Transient effects are discussed in Chapter 13.

9.1.1 Ionization effects

9.1.1.1 Basic considerations

The first process we will consider is ionization. For charged particles the most likely interaction is with electrons in the target material because electrons occupy most of the atomic volume. In this process, referred to as ionization loss, a small amount of energy – typically a few eV – is transferred to an electron by the incoming particle. For a semiconductor or insulator, this energy is absorbed by an electron in the valence band, raising it to the conduction band, and creating a corresponding hole in the valence band. Thus, ionization creates an electron-hole pair in the material. This is shown pictorially in the E-k diagram of Fig. 9-1.

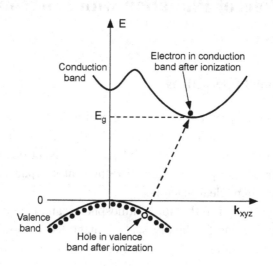

Fig. 9-1. Creation of an electron-hole pair by ionization in a semiconductor or insulator.

When ionization is produced by photons, the minimum energy required to create an electron-hole pair is the bandgap energy, E_g, because the photon is directly absorbed by the valence band electron. However, for high energy particles the process is more complicated. Ionization from a heavy particle is the result of a shower of secondary electrons, which have a mean free path that overlaps several lattice positions.

Part of the deposited energy is dissipated through other processes, including weak interactions with other electrons that do not result in ionization. For this reason the minimum amount of absorbed energy required to create an electron-hole pair from high-energy particles is larger than for photon absorption. The minimum energy depends on the particle type and energy. It is typically between 2.2 and 4 times the bandgap energy [1,2]. Table 9-1 shows the bandgap and minimum energy for electron-hole creation through ionization for several semiconductors, as well as SiO_2.

Table 9-1. Bandgap and Energy for Electron-Hole Pair Production by Ionization for Several Materials.

Material	E_G (Ev)	Minimum Energy to Produce an Electron-Hole Pair (eV)
Si	1.12	3.6
InP	1.35	4.5
GaAs	1.43	4.7
SiC	2.86	8.5
SiO_2	9.1	18

The unit used to describe absorbed dose due to ionization is the rad, defined as 100 ergs absorbed in 1 g of a specific material. The S.I. unit for absorbed dose is the gray, defined as 1 j/kg (1 gray = 100 rads), but the rad is more commonly used. The material in which the dose is deposited needs to be included for these units, *e.g.*, 1 rad(GaAs) for deposited dose in GaAs, because the mass absorption coefficient is not the same for different materials.

As discussed in the previous chapter, total dose levels in typical space applications range from less than 5 krad(Si) for the International Space Station to as much as 1-Mrad(Si) for a spacecraft that goes through the intense regions of the earth's radiation belts, or in the radiation belts near Jupiter.

9.1.1.2 Effects in insulators

Ionization takes place in insulating regions, such as the high-quality oxides used in silicon semiconductor devices, as well as in the semiconductor regions. Although insulators are poor conductors, the charge from ionization usually migrates from its initial location after it is produced. For most insulators electrons have sufficiently high mobility that they will be swept out even if only a weak electric field is present. Holes have lower mobility, and may either remain in place (at temperatures less than approximately 100 K for silicon dioxide), or gradually migrate at normal temperatures. The migration process depends on the properties of the material as well as the magnitude of any electric field and the competing process of electron-hole recombination.

Charge transport within semiconductors and insulators depends on the electric field. Charge traps are formed at (or near) interface regions after the transport process is completed. Although electron trapping can occur, hole traps usually dominate. Hole transport can be extremely slow in insulators where only a weak electric field is present.

The direction of the electric field determines the direction for hole movement (holes migrate from a positive potential towards more negative potentials). If the holes migrate towards contacts they will have less effect on device properties than if they migrate towards the semiconductor-insulator interface. Some types of traps anneal with time, even at room temperature, causing radiation-induced changes in device parameters to be time dependent as well as field dependent.

Absorbed dose is the total amount of energy that goes *into* ionization loss processes. The net effect in a semiconductor device depends on the way that excess charge produced by the absorbed dose reaches an equilibrium condition within the device structure after the ionization process takes place. In nearly all cases some of the excess charge will recombine before it can become trapped, reducing the effect of the absorbed dose when one considers the net effect on device parameters.

Semiconductor devices that are sensitive to charge trapping – particularly MOS transistors – are strongly affected by ionization damage. The traps change key device properties such as the gate threshold voltage and leakage current. Similar effects can take place at surfaces in compound semiconductors, but are usually of less importance compared to silicon because of the higher surface state density in those materials (the consequence of the higher surface state density is that much higher total dose levels are required to affect the surface state density). Many compound semiconductors use buffer or capping layers that isolate surface trapping from the active regions of the device.

Therefore, only an abbreviated treatment will be given to ionization damage in this book. An excellent reference for ionization effects is the book edited by Ma and Dressendorfer [3].

9.1.2 Particle scattering

The second mechanism is displacement damage, which moves lattice atoms away from their normal position, disrupting the regular atomic spacing within the material. Before we discuss displacement damage, we need to briefly review some important results for particle scattering. Particle collisions can be divided into two general categories: elastic scattering, where the total energy of the incident particle and target is unchanged by the scattering event; and inelastic scattering, where additional energy is lost or gained during the collision through other processes, such as nuclear reactions.

Many interactions of space radiation can be understood from the standpoint of Rutherford scattering, which describes scattering of two charged particles. The primary particle has charge $Z_1 e$ and momentum $m_1 v_0$. That particle interacts through electrostatic forces (Coulomb scattering), with a target particle of mass m_2 and charge $Z_2 e$. The extrapolated direction of the initial path of the incoming particle (without considering the change in angle from scattering) comes within a distance b of the initial site of the target particle, as shown in Fig. 9-2. The distance b is called the impact parameter. For the range of energies encountered by protons in space, nonrelativistic scattering can be used, but relativistic scattering must be used for electrons.

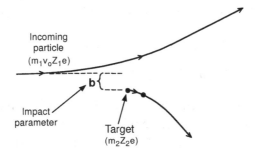

Fig. 9-2. Diagram of the Rutherford scattering process.

The distribution of scattering angles of the incoming particle is inversely proportional to the fourth power of the sine of ½ the scattering angle. Consequently only a small fraction of the particles are scattered at large angles (an angle of 180° corresponds to backscattering), provided that the mass of the incoming particle is less than that of the target particle. That condition corresponds to maximum energy transfer from the incoming particle to the target particle.

For the nonrelativistic case, the maximum energy that can be transferred in an elastic collision is

$$E_{max} = 4E_{inc}\left[\frac{m_1 m_2}{(m_1 + m_2)^2}\right] \qquad (9\text{-}1)$$

where E_{max} is the maximum energy that can be transferred, E_{inc} is the energy of the incoming particle, m_1 is the mass of the incident particle, and m_2 is the mass of the target atom.

When only a small amount of energy is lost by the incoming particle, its velocity and direction are nearly unaffected by the collision process, and we can treat the collision as imparting a slight impulse of momentum to the target particle, perpendicular to the direction of motion of the incoming particle. This is described by

$$p = \frac{Z_1 Z_2 e^2}{2\pi \varepsilon_0 v_0 b} \qquad (9\text{-}2)$$

where p is the momentum impulse, ε_0 is permittivity, and the other parameters are described in Fig. 9-1. The impulse is inversely proportional to the speed of the incoming particle because it spends more time in the vicinity of the target particle. It is also inversely proportional to the impact parameter.

9.1.3 Displacement damage

Displacement damage occurs where an incident particle transfers enough energy to move the target atom from its normal lattice position to another position, creating a vacancy in the lattice. The interaction can either be electromagnetic (for charged particles with lower energies) or through nuclear interactions. The cross section for nuclear mechanisms is much lower than for ionization because the effective cross section ranges from the Bohr radius to the size of the nucleus, both of which are small compared to ionization cross sections that involve electrons and holes in the extended crystal lattice.

The threshold energy, which is the minimum energy that must be absorbed by a lattice atom to move it from its normal position, is closely related to crystal binding energy, and is roughly correlated with the reciprocal of the lattice spacing. The threshold energy is approximately 10 eV for GaAs, and about 21 eV for SiC. Table 9-2 shows the displacement threshold energy for several semiconductors. In general, semiconductors with high threshold energy will be more resistant to displacement damage because of the higher threshold energy.

Table 9-2. Displacement Threshold Energy for Several Semiconductors.

MATERIAL	DISPLACEMENT ENERGY THRESHOLD (eV)
InAs	7.4
InP	7.8
GaAs	9.5
Si	12.9
Ge	14.5
GaN	19.5
4H-SiC	21.3

9.1.3.1 Electron displacement damage

Scattering of high-energy electrons also obeys the Rutherford scattering law, but the electrons in space environment have sufficiently high velocity that the relativistic form of the Rutherford cross section must be used. For relativistic particles, the maximum energy that can be transferred to a nucleus of mass M by a head-on collision is described by

$$E_m = \frac{2mE}{M}\left[\frac{2+E/mc^2}{(1+m/M)^2(Mc^2)+2E}\right] \quad (9\text{-}3)$$

where E_m is the maximum energy that can be transferred, E is the energy of the incident particle, and m is the mass of the electron. Thus, there is a distinct threshold energy for electron scattering.

The threshold energy for a specific material can be determined by measuring the energy dependence of displacement damage. Figure 9-3 shows experimental results for GaN, with a threshold energy of 440 keV [4]. From Eq. 9-1, that corresponds to a displacement energy threshold of 19 eV.

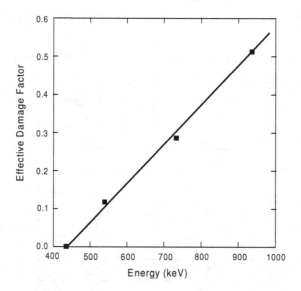

Fig. 9-3. Plot of effective damage factor vs. electron energy for GaN showing the threshold energy [4]. © 2002 IEEE. Reprinted with permission.

A wide range of energies can be transferred to the lattice atom by the mechanisms that produce displacements. For the case where a relatively small amount of energy is absorbed – that is, energies of the same order as the threshold energy – a vacancy-interstitial pair (Frenkel pair) is formed where the vacancy and interstitial atom are located close together. This is illustrated in Fig. 9-4.

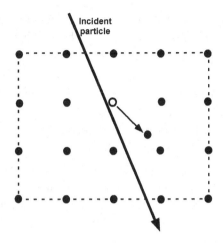

Fig. 9-4. Diagram of displacement of a lattice atom forming a vacancy-interstitial pair.

The microscopic nature of the damage is quite different for more energetic collisions, where the energy absorbed by the lattice atom is much higher than the threshold energy. The atom involved in the primary interaction with the incident radiation particle is called the *primary knock-on atom* (PKA). If the PKA energy is much greater than the displacement energy threshold, it will have sufficient energy to interact with several other atoms in the lattice before it stops, creating additional atomic displacements.

9.1.3.2 Displacement damage from protons and heavy ions

Interactions of high energy particles depend on the mass and energy of the incoming particle as well as the mass of the target atom. Displacement damage for more massive charged particles can be produced when their velocity is well below the speed of light.

For an incident charged particle with low energy (compared to its rest mass), the transit time is long enough so that the more tightly bound electrons in the target atom can move in response to the electric field from the particle strike, increasing the effective size of the nucleus in the collision process.

Collisions in that energy range are approximately equivalent to "hard sphere" collisions with a radius of $a_o Z^{-1/3}$, where a_o is the Bohr radius and Z is the atomic number. Note that this value is much larger than the actual size of the nucleus.

For particles with higher energies, the effect of the screening electrons is reduced – they cannot respond quickly enough for high velocity particles – which in turn reduces the effective collision radius. As a result the cross section is lower for higher energy particles when we deal with more massive particles, such as protons or heavy ions.

For hard sphere collisions the number of displaced atoms produced by a secondary PKA particle with energy $\hat{E}$ can be estimated from the displacement threshold energy, E_d as follows [5]:

$$N_d = 1 \text{ for } 0 < \hat{E} < 2 E_d,$$

$$N_d = \frac{\hat{E}}{2E_d} \text{ for } 2E_d < \hat{E} < E_i$$

(9-4)

E_i is the energy of the primary particle that produces the PKA. For example, a 1-keV recoil atom in GaAs (E_d = 9.5 eV) can produce up to 50 Frenkel pairs.

These relationships only apply to simple processes where a series of Frenkel pairs are produced by the initial PKA atom as it traverses the lattice. PKA atoms with higher energy produce more complicated damaged regions, including cascades (or clusters), where an extended region of the lattice is damaged by the recoil atom as it gradually dissipates excess energy.

Figure 9-5 shows the results of a Monte Carlo calculation of displacement damage in silicon for the case where 50 keV of energy – about 4000 times greater than the displacement threshold energy – is transferred to the PKA [6] using a more sophisticated model that accounts for cluster defects. In this example, the atom follows a tortuous path through the lattice, creating two basic types of damage: vacancy-interstitial pairs, and large *cascade damage* regions, as shown in the figure. The displaced lattice atom finally stops about 70 nm away from its original position, a distance corresponding to several hundred lattice sites from its original position.

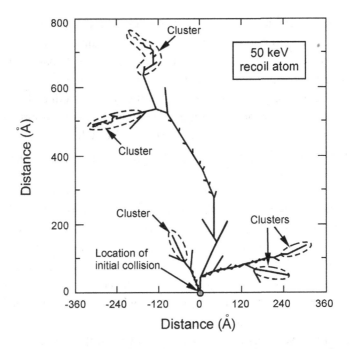

Fig. 9-5. Displacement damage in silicon for the case where the energy transferred to the lattice site is more than 1000 times greater than the threshold energy for displacement [6]. © 1972 IEEE. Reprinted with permission.

Although this calculation is for silicon, the same process occurs in other materials. The important point is that the microscopic nature of the damaged region is very different for the case where high amounts of energy are absorbed by the lattice atom. The cascade damage regions extend over many lattice sites, and are charged after they are formed, as well as disrupting the crystal lattice over an extended region. They are usually unstable, except at low temperature. Part of the damage may recover after irradiation (annealing). Annealing in GaAs – as well as for most other materials - depends on charge injection after the damage has occurred. Thus, passing current through an irradiated device will accelerate the annealing process.

The difference in microscopic damage for various particles and energies is extremely important. Incident particles with low energy produce Frenkel pairs. Particles with higher energy also produce Frenkel pairs, but also create cascade regions where there are many displaced atoms in a localized region. The approximate amount of energy that must be transferred to a lattice atom to produce cascade damage regions is 1000 eV. For the case where higher amounts of energy are absorbed by the lattice atom, very complex damage patterns are produced, as shown in Fig. 9-5. Such complicated damage regions become important when the lattice atom absorbs an energy $>10^4$ eV. We will return to this point later, in the discussion of displacement damage from various types of primary particles and energies.

The complex lattice structure of compound semiconductors makes it more difficult to analyze displacement processes compared to a material with only one type of atom. For a binary compound such as GaAs the threshold energy is different for the two atomic species because they have different crystal binding energies. The range of the displaced atom and its effect on the lattice also depends on the collision angle. However, those distinctions are usually unimportant when we consider the effects of many such collisions on macroscopic device properties unless we are dealing with low energy particles that have barely enough energy to displace individual atoms.

9.2 Effects of Damage on Semiconductor Properties

9.2.1 Lifetime damage

Early radiation effects studies noted that the semiconductor parameter most affected by displacement damage was minority carrier lifetime, which becomes shorter after damage occurs.

Studies of basic materials showed that the minority carrier lifetime could be related to particle fluence by the equation [7]

$$\frac{1}{\tau} - \frac{1}{\tau_o} = \frac{\Phi}{K_{tau}} \qquad (9\text{-}5)$$

where τ is the minority carrier lifetime after irradiation, τ_o is the initial minority carrier lifetime, Φ is the particle fluence (#/cm^2), and K_{tau} is the lifetime damage constant (cm^{-2}-s per particle). The damage constant,

K_{tau}, depends on particle type, particle energy, (and for some semiconductors the injection level), and is really only constant under a restricted set of conditions.

Lifetime degradation can often be related to electronic device properties. One example is bipolar transistor gain for transistor technologies for the case where gain is limited by recombination in the base region. Changes in lifetime affect the minority carrier diffusion length L

$$L = \sqrt{D\tau} \qquad (9\text{-}6)$$

where D is the diffusion constant. The diffusion length is related to transistor gain through the equation [8]

$$\frac{1}{h_{FE}} - \frac{1}{h_{FEO}} = \frac{\Phi}{2\pi f_T K_{tau}} \qquad (9\text{-}7)$$

where h_{FEO} is the initial gain before irradiation, h_{FE} is the reduced gain after irradiation, f_T is the gain-bandwidth product of the transistor, and K_{tau} is the lifetime damage constant in the previous equation (9-2). Equation 9-7 is widely used for silicon transistors, but is not necessarily applicable to transistors with very thin base regions, such as HBTs. Nevertheless, it is important to understand Eq. 9-7 because a great deal of the literature on displacement damage is compared to transistor gain degradation in silicon. Silicon is frequently used as a benchmark, even when discussing displacement damage in compound semiconductors.

Lifetime damage also affects solar cells and photodetectors, which often depend on charge collection by diffusion. Lifetime damage is of secondary importance for JFETs and MESFETs, because they are majority carrier devices. In fact, relatively few studies of lifetime damage have been done for compound semiconductors. Later we will infer the effects of radiation on minority carrier lifetime by examining specific devices – primarily light-emitting diodes and optical detectors – where minority carrier lifetime is important in their overall performance.

The literature on displacement damage is highly inconsistent in the way that the term damage constant is defined and applied. A different form of the lifetime damage constant, the reciprocal of the damage constant K_{tau} in Eqs. 9-2 and 9-3, is generally used to describe damage in solar cells, as well as in some studies of transistors [9]. Unfortunately the same symbol, K, is used, adding to the confusion. For that reason we have used the subscript tau for the older and more fundamental lifetime damage constant in those equations.

In many device studies parameters related to lifetime are either unavailable or simply not measured. This has lead to adoption of a damage *factor* that lumps the lifetime and damage constant together. The symbol K is often used for the damage factor, particularly in compound semiconductors, resulting in a very confusing situation: K is used in three different ways in the literature. In this book we will use the symbol C for the damage factor, and retain the subscript tau in the lifetime damage constant symbol. Equation 9-7 shows how transistor gain degradation can be described in this way:

$$\frac{1}{h_{FE}} - \frac{1}{h_{FEO}} = C\Phi = \frac{1}{K_{tau}\tau} \qquad (9\text{-}8)$$

Relatively few lifetime damage studies have been done at the device level for compound semiconductors. Nevertheless, it is useful to compare typical values for various materials. Table 9-3 compares lifetime damage constants for 50-MeV protons in a silicon transistor with damage constants for compound semiconductor LEDs. Note the much higher values for the two widegap semiconductors, which causes them to be much less sensitive to radiation damage.

Before leaving this topic we need to note that the damage constant K_{tau} has units of cm^2-s/particle, and does not depend directly on lifetime (although it may depend on lifetime for other reasons), while the damage factor, C, explicitly depends on lifetime, even though lifetime is usually not measured when the damage factor is employed. Thus, we expect a much wider range in values of the damage factor compared to values of the damage constant because of the wide range of minority carrier lifetimes that are encountered in different samples of a specific semiconductor.

Table 9-3. Lifetime Damage Constants for 50-MeV Protons in Various Materials.

Device Structure	Material	Approximate Doping Level (#/cm³)	Damage Constant [cm²-s/p]
Silicon Transistor	P-Silicon *(base region)*	5×10^{17}	$\sim 7 \times 10^5$
GaAs LED	n-GaAs (Si doped)	$\sim 10^{19}$ (compensated)	$\sim 3 \times 10^4$
GaN LED	undoped	unspecified	$\sim 10^7$
6H SiC	unspecified	unspecified	$\sim 4 \times 10^7$

9.2.2 Carrier removal

A second mechanism, which is most often the dominant damage mechanism for compound semiconductors, is carrier removal. This process occurs because radiation-induced defects within the bandgap can trap some of the majority carriers from dopant atoms that are normally ionized to produce carriers within the valence or conduction band. This effectively changes the doping density as the number of defects increases during irradiation. The basic relation (shown here for an n-type semiconductor) is

$$n - n_o = R_c \Phi \qquad (9\text{-}9)$$

where n is the minority carrier density, n_o is the initial minority carrier density before irradiation, R_c is the carrier removal coefficient in cm⁻³ per particle/cm², and Φ is the particle fluence. In the literature the units are often simplified to cm⁻¹, but that is not really correct because we are comparing the density of dopant atoms with the fluence of incoming particles, which are generally different species.

Before we discuss carrier removal in more detail, it is necessary to examine the effect of large numbers of defects on a semiconductor. If the number of defects is small, the Fermi level will remain unchanged. However, as the number of defects increases they will affect the bandgap, depending on the position of the recombination centers in the bandgap relative to the energy of the dopant atoms. Defect centers that are located deep within the bandgap are generally more effective in the carrier removal process compared to shallow defects, and the carriers that are captured are more thermally stable compared to shallow defects.

Dopant atoms are normally close to either the valence or conduction band, but the recombination centers can occur throughout the forbidden region. As the bandgap changes, the linear relationship implied by Eq. 9-9 will no longer apply, causing the carrier removal coefficient to change with fluence. R_c is usually constant as long as the change in carrier density is <25%, but for compound semiconductors we are often concerned with fluences that are high enough to produce larger carrier densities. Williams, *et al.*, developed an exponential fit that allows the recombination coefficient to be extended to higher radiation levels for GaAs [10].

Older data on carrier removal can be confusing because of different ways of measuring and interpreting results. A more recent analysis [11] considered two different components for carrier removal: one involving deep traps, that is unaffected by small changes in the Fermi level, and one involving shallow traps – typically within an energy of ~kT of the Fermi level – that is highly sensitive to the position of the Fermi level. Using this approach, the deep trap component in GaAs was found to be independent of doping level, with an approximate value of 3 cm^{-1}, even though the total carrier removal rate increased with doping density (only the shallow trap component was affected by the doping density). This may not be the case for other compound semiconductors, but appears to be valid for GaAs.

The data of Williams, *et al.*, are in reasonable agreement with older results for Gunn diodes from Marcus and Breummer [12], who developed an empirical relationship for the dependence of carrier removal rate on doping density

$$R_c = 9.08 \times 10^{-6} N_o^{0.4} \qquad (9\text{-}10)$$

where N_o is the pre-irradiation value of the doping density. Although the recombination rate increases with N, it is sub-linear, causing the relative change in doping concentration at a given fluence to be smaller as doping density increases. A similar relationship has been developed for silicon, with a higher exponent, (0.77) [13].

Other compound semiconductors have not been studied as extensively as GaAs. Messenger, *et al.*, determined that carrier removal in InP was independent of carrier concentration over the range of 10^{16} to 4×10^{17} cm^{-3}, using solar cell data [14]. McLean, *et al.*, established a carrier removal rate for SiC JFETs that was approximately 5 cm^{-1}, using neutron damage [15].

The carrier removal rate is effectively a damage constant (with different dimensions) that applies to the carrier removal process. Unlike the lifetime damage constant, which varies over a wide range for different materials (see Table 9-3), carrier removal rates are roughly the same for different materials. For example, the carrier removal rate for SiC is only about three times lower than that of silicon, even though the binding energy of SiC is much higher [16]. Part of the reason is that only some of the defects created by the displacement damage process affect the carrier density through the carrier removal mechanism.

9.2.3 Mobility

Mobility is also affected by displacement damage because impurities within the bandgap increase carrier scattering. Mobility changes can be described by

$$\mu_o/\mu = 1 + \beta\Phi \qquad (9\text{-}11)$$

where β is the damage constant that applies to this process. Damage constants for mobility are lower than for lifetime damage or carrier removal, and consequently the effect of displacement damage on mobility is usually only important at very high fluences.

McGarrity, *et al.*, developed a relationship between carrier removal and the mobility damage constant that is applicable when the damage is low enough to prevent the Fermi level from changing [16]. It starts with the assumption that the carrier removal rate is equal to the introduction rate of recombination centers, allowing the mobility to be expressed as

$$\mu = \mu_o \left(1 + \chi \frac{\Delta n_o}{n_o}\right)^{-1} \quad (9\text{-}12)$$

where μ is the mobility after irradiation, μ_o is the initial mobility, Δn_o is the change in carrier concentration, n_o is the initial carrier concentration, and the parameter χ is defined as

$$\chi = \frac{\mu_o n}{\alpha} \quad (9\text{-}13)$$

with α the coefficient that describes the (linear) relationship between the change in mobility and the change in mobility due to Coulomb scattering from charge centers.

With further assumptions, they obtain the equation

$$\mu = \mu_o \left[1 - \frac{\chi}{1+\chi}(1 - f_\phi)\right] \quad (9\text{-}14)$$

where f_ϕ, a parameter that is slightly less than unity, is related to experimentally related quantities for the JFETs in their study. When they applied this equation, they found that the parameter χ was 0.6, showing that we expect changes in mobility to occur at fluences that are only slightly above the fluences where carrier removal starts to affect the carrier concentration.

Figure 9-6 shows how minority carrier lifetime, carrier density, and mobility of a GaAs LED depend on proton fluence (all three are normalized to their pre-irradiation values). Initial values were as follows: lifetime, 0.9 μs; carrier concentration 3×10^{16} cm^{-3}; and mobility, 6500 cm^2/(V-s). The lifetime results are direct measurements, while the values for carrier concentration and mobility are calculated. Note the extreme sensitivity of minority carrier lifetime to radiation damage compared to the other parameters.

If the initial lifetime value was an order of magnitude lower (0.09 μs), the fluence dependence for liftetime damage would shift by one order of magnitude to right, while the fluences for changes in carrier concentration and mobility would be about the same.

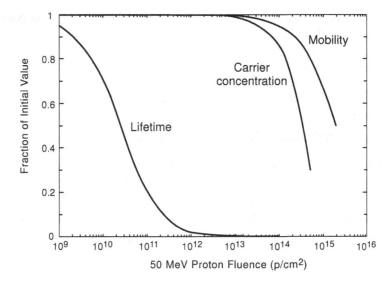

Fig. 9-6. Degradation of minority carrier lifetime, carrier concentration, and mobility for a GaAs light-emitting diode after irradiation with 50 MeV protons.

This illustrates why the few types of compound semiconductors that require long lifetimes for their electrical performance are highly sensitive to displacement damage, while the majority of compound semiconductor devices are relatively resistant to displacement damage effects. Note that the fluence range in Fig. 9-6 between lifetime damage and changes in carrier concentration (or mobility) is about five orders of magnitude!

9.3 Radiation Effects in Heterostructures

Heterostructures play an important role in many compound semiconductors. The most important mechanism for damage in heterostructures is usually carrier removal, but we have to deal with carrier removal effects in several different regions, with different doping concentrations and materials. The change in doping concentration can alter the bandgap of the two materials, particularly the low doped (or undoped) region that is often used as the active layer in order to take advantage of the high electron mobility for compound semiconductors with low doping concentration. We will refer details to the next chapter, which discusses radiation effects in specific types of compound semiconductor devices.

Some of the effects that can occur in heterostructures include:

- Increase in tunneling across the thin heterostructure barriers,
- Changes in the sheet charge that affects the carrier density within the 2DEG region, and
- Changes in the energy levels within quantum well structures.

Most of these mechanisms are only important at very high fluence levels, as expected from the previous discussion. Nevertheless, it is important to understand the way that they affect device performance, including the assumption of a linear relationship between particle fluence and parameter changes that is the usual starting point for analyzing device damage effects.

9.4 Energy Dependence of Displacement Damage

As discussed in the previous chapter, the space environment consists of various types of particles with different energies. We need to know the energy dependence and relative effects of these different particles on devices in order to determine how the devices are affected by space radiation. There is also considerable interest in the effects of neutrons on devices, partly because neutrons are a convenient radiation source for displacement damage studies. With sufficient understanding of the mechanisms and energy dependence it is possible to apply experimental results for one set of particle types and energies to a different set. This has been a long-time goal of the research community.

Experiments and theoretical calculations have been used for many years to determine how particle type and energy affect radiation damage. A great deal of progress has been made, but just as for ionization damage, we have to carefully distinguish between the total energy going into displacement processes (which is more straightforward) and the net effect on a device after redistribution and annealing. Charge injection may also affect damage recovery, adding a further level of complexity to the problem.

9.4.1 Displacement energy comparisons

Current practice is to use the *non-ionizing energy loss* (NIEL) concept to describe displacement damage effects in materials [17]. The basic idea is to lump non-ionizing processes together, and assume that a displacement damage dose (analogous to ionizing radiation dose) is deposited in the material from those interactions. The units of displacement dose are MeV/g (normalized to the material density). Although the NIEL concept is appropriate from the standpoint of partitioning the deposited energy, it does not consider the net damaging effect in a semiconductor, which depends on how the damaged region stabilizes afterwards.

An example of the results of NIEL calculations for GaAs is shown in Fig. 9-7, based on values from References 17 and 19. Proton damage follows a 1/E dependence for energies below 10 MeV. As discussed earlier, this occurs because protons with lower velocity spend more time in the vicinity of the nucleus, increasing the impact parameter for electromagnetic interactions, the dominant mechanism at low energy.

More complex nuclear reactions become important at higher energies, and consequently the energy dependence departs from the 1/E dependence at higher energies.

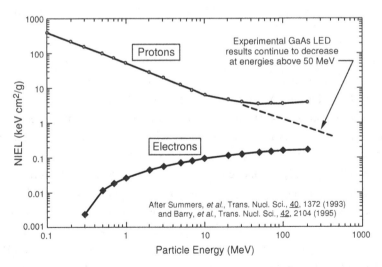

Fig. 9-7. Non-ionizing energy calculations for protons and electrons in GaAs [17,19]. © 1993 and 1995 IEEE. Reprinted with permission.

Displacement damage calculations are more straightforward for a semiconductor such as silicon with a single element, compared to compound semiconductors. Initial attempts to calculate NIEL for GaAs were less successful than for silicon, requiring several model corrections before reasonable agreement could be obtained with experimental results [18], and there are still disparities at high energies that have not been fully resolved.

NIEL calculations for other compound semiconductors have very nearly the same dependence. Table 9-4 compares NIEL values for silicon, GaAs and InP for four different proton energies [17]. There are slight differences, but the results are about the same – within approximately 20% - except at very high energies.

Table 9-4. Proton NIEL Values (MeV/g) for Various Materials at Different Energies.

MATERIAL	E = 1 MeV	E = 10 MeV	E = 50 MeV	E = 200 MeV
Si	6.73×10^{-2}	7.88×10^{-3}	3.88×10^{-3}	1.94×10^{-3}
GaAs	5.40×10^{-2}	6.59×10^{-3}	3.75×10^{-3}	3.93×10^{-4}
InP	5.53×10^{-2}	7.24×10^{-3}	3.36×10^{-3}	3.54×10^{-4}

Electron results are also similar for different materials. The main difference is the sharp threshold energy, which depends on the binding energy of the lattice. As shown earlier in Fig. 9-3, the threshold energy for GaN is about 440 keV. The minimum electron energy for displacement in GaAs is about 250 keV, which is roughly correlated with the difference in binding energy of the two materials.

The NIEL concept is extremely useful. In many cases it agrees very well with experimental results for different types of particles and energies. However, it does not take microscopic damage into account. Most of the papers that deal with NIEL calculations clearly point that out. The main purpose of NIEL is to evaluate the relative damage effect of different particles, using experimental values to determine the damage constant, and simplifying the application of experimental results at a few energies with the continuous spectrum of energies that are found in space environments.

9.4.2 Discrepancies between NIEL and experiments

It is hardly surprising that there are discrepancies between NIEL calculations and experiments on device damage, particularly because of the differences in microscopic damage for low and high energy interactions. For protons there is generally good agreement for energies <10 MeV, where hard-sphere collisions are the dominant process. Most of the discrepancies arise at higher energies, where nuclear collisions are involved. Extensive work has been done for silicon, and there is satisfactory agreement between the energy dependence based on NIEL calculations and experimental damage for that material.

The situation is more complicated for GaAs due to two factors: first, because two atoms are involved, there are more possible interactions as well as complicated angular dependencies; and second, damage in GaAs devices usually depends on carrier removal, not minority carrier lifetime. Shallow impurities are more effective in producing lifetime damage compared to carrier removal, which is one possible reason for the observed discrepancies.

Lifetime degradation is the dominant mechanism for displacement damage in most types of LEDs. Figure 9-8 compares calculations of NIEL – which agree reasonably well with carrier removal damage in GaAs JFETs – and experimental values of lifetime damage in GaAs LEDs [19]. Two calculated values are shown, one corresponding to ealier studies, and a second, labeled "Restricted Energy Loss" that includes revisions to the model that attempted to resolve differences between theory and experiment for carrier removal. There is good agreement at low energies, but a large discrepancy is evident at higher energies where inelastic collisions dominate the interaction. For 200 MeV protons, the discrepancy between the initial calculations and experimental values is about a factor of six. The improved calculations reduced the error to about a factor of two, but larger errors are present even for the restricted energy loss model at higher energies.

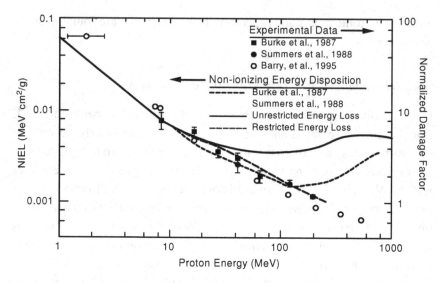

Fig. 9-8. Experimental results for the energy dependence of GaAs LEDs compared with NIEL calculations in GaAs [19]. © 1996 IEEE. Reprinted with permission.

For space environments this discrepancy is relatively unimportant because high energy protons comprise only a small fraction of the energy spectrum. However, large errors can occur if radiation tests are done at proton energies in the region where the disparity occurs, and the equivalent damage at lower proton energies is calculated from older studies that do not apply to the actual devices that are being considered. Many radiation tests have been done using 200-MeV protons (or 1 MeV neutrons) because of experimental convenience, including the long range of the particles which allows one to do the tests without correcting the results for energy loss through the device package or through top layers of the structure above the active layer.

A similar discrepancy exists for GaN LEDs at higher energies [20].

9.4.3 Annealing

Annealing occurs in defects from displacement damage, and this is one of the most important factors in interpreting radiation damage in devices. Annealing may occur spontaneously, depending on temperature, or it may be accelerated by either current or photons (injection-dependent annealing).

For devices that are unbiased, isochronal annealing can be used to determine how temperature affects annealing. Silicon anneals at room temperature, but displacement damage in GaAs is effectively stable for temperatures below 200°C.

Although GaAs defects are stable for unbiased devices, GaAs is strongly affected by injection-enhanced annealing, even at room temperature [21]. InGaP also exhibits injection-enhanced annealing [21]. Annealing can proceed very rapidly during the process of measuring device properties after irradiation, and it is extremely important to account for this effect when devices are measured after irradiation. Even though it is an annoying complication for device measurements, it is also possible to take advantage of injection-enhanced annealing to allow degraded devices to recover in space.

Annealing in photonic devices is affected by optical power as well as carrier injection. We will defer a detailed discussion of annealing to Chapter 11, which discussed displacement damage in optoelectronic devices.

9.5 Summary

This chapter has discussed the fundamental interactions of charged particles with semiconductors. Elementary scattering theory explains why the effective damage for protons increases at low energy, while the damage for electrons has a sharp threshold energy and a flatter energy dependence at high energy. We have to understand the way that displacement damage depends on particle type and energy in order to apply experimental results that are limited in scope to only a few conditions to the continuous range of energies that are encountered in space environments.

The concept of non-ionizing energy loss (NIEL) has been widely adopted as a way to compare damage for different types of particles and particle energies. It is a very useful concept, but it does not incorporate annealing effects, and is not completely successful at high particle energies. Part of the reason for this is the difference in the microscopic damage that is produced when different amounts of energy are absorbed from an incident particle by a crystal atom. For low energies, lattice vacancies are close to the interstitial atom produced by the collision, whereas collisions with high energy transfer produce a very complicated damage region within the material that can extend for hundreds of lattice sites. Despite these differences, the NIEL concept is widely used. It can be very effective in interpreting damage for the space environment, but care must be taken to ensure that experimental results are not entirely based on the high-energy region where discrepancies between NIEL can be significant.

Fundamental considerations show that most compound semiconductors are highly resistant to displacement damage (as well as ionization damage), primarily because they are majority carrier devices that are relatively insensitive to minority carrier lifetime damage. However, minority carrier lifetime is an important parameter for most optoelectronic devices, and significant changes in minority carrier lifetime can occur at very low radiation levels. Some categories of optoelectronics are so strongly affected by displacement damage that they may fail within 1–2 years in space applications.

References

1. C. Bussolati and A. Fiorentini, "Energy for Electron-Hole Pair Generation in Silicon by Electrons and Protons", Phys. Rev., **136**(6A), pp. A1756–A1758 (1964).
2. G. A. Ausman, Jr., and F. B. McLean, "Electron-Hole Pair Creation Energy in SiO_2", Appl. Phys. Lett., **26**(4), pp. 173–175 (1975).
3. T. P. Ma and P. V. Dressendorfer, *Ionizing Radiation Effects in MOS Devices and Circuits*, John Wiley: New York (1989).
4. A. Ionascut-Nedelcescu, *et al.*, "Radiation Hardness of Gallium Nitride", IEEE Trans. Nucl. Sci., **49**(6), pp. 2733–2738 (2002).
5. G. H. Kinchin and R. S. Pease, "The Displacement of Atoms in Solids by Radiation", Rept. On Progress in Physics, **18**(1), pp. 1–54 (1955).
6. V. A. J. Van Lint, R. E. Leadon and J. F. Colwell, "Energy Dependence of Displacement Effects in Semiconductors", IEEE Trans. Nucl. Sci., **19**(6), pp. 181–185 (1972).

7. O. L. Curtis, et al., "Effect of Irradiation on the Hole Lifetime of n-Type Germanium", J. Appl. Phys., **28**, pp. 1161–1165, Oct., 1957.
8. G. C. Messenger and J. P. Spratt, "The Effects of Neutron Irradiation on Germanium and Silicon", Proc. of the IRE, pp. 1038–1044, June, 1958.
9. J. R. Srour, C. J. Marshall and P. J. Marshall, "Review of Displacement Damage Effects in Silicon Devices", IEEE Trans. Nucl. Sci., **50**(3), pp. 653–670 (2003).
10. J. G. Williams, et al., "Carrier Removal and Changes in Electrical Properties of Neutron Irradiated GaAs", J. Appl. Phys., **70**(9), pp. 4931–4937 (1991).
11. S. T. Lai, D. Alexiev and B. D. Nener, "Comparison between Deep Level Defects in GaAs Introduced by Gamma, 1 MeV Electron, and Neutron Irradiation", J. Appl. Phys., **78**(6), pp. 3686–3690 (1995).
12. G. E. Marcus and H. P. Bruemmer, "Radiation Damage in GaAs Gunn Diodes", IEEE Trans. Nucl. Sci., **17**(6), pp. 230–232, Dec. 1970.
13. B. Buchanan, R. Dolan and S. Roosild, "Comparison of the Neutron Radiation Tolerance of Bipolar and Junction Field Effect Transistors", Proc. of the IEEE, pp. 2188–2189 (1967).
14. S. R. Messenger, et al., "Carrier Removal in p-Type InP", IEEE Trans. Nucl. Sci., **45**(6), pp. 2857–2860 (1998).
15. F. B. McLean, et al., "Analysis of Neutron Damage in High-Temperature Silicon Carbide JFETs", IEEE Trans. Nucl. Sci., **41**(6), pp. 1884–1894 (1994).
16. J. M. McGarrity, et al., "Silicon Carbide JFET Radiation Response", IEEE Trans. Nucl. Sci., **39**(6), pp. 1974–1981 (1992).
17. G. P. Summers, et al., "Damage Correlations in Semiconductors Exposed to Gamma, Electron and Proton Radiations", IEEE Trans. Nucl. Sci., **40**(6), pp. 1372–1379 (1993).
18. G. P. Summers, et al., "Energy Dependence of Proton-Induced Displacement Damage in Gallium Arsenide", IEEE Trans. Nucl. Sci., **34**(6), pp. 1220–1226 (1988).
19. A. L. Barry, et al., "The Energy Dependence of Lifetime Damage Constants in GaAs LEDs for 1–500 MeV Protons", IEEE Trans. Nucl. Sci., **42**(6), pp. 2104–2107 (1995).
20. S. M. Khanna, et al., "Proton Energy Dependence of the Light Output in Gallium Nitride Light-Emitting Diodes", IEEE Trans. Nucl. Sci., **51**(6), pp. 2729–2735 (2004).

21. R. Loo, R. C. Knechtli and G. S. Kamath, "Enhanced Annealing of GaAs Solar Cell Damage", in *Proceedings of the 15th Photovoltaic Specialties Conference*, Vo. 33 (1981).
22. M. Yamaguchi, *et al.*, "Radiation Resistance of InGaP Solar Cells", in *Proceedings of the 25th Photovoltaic Specialties Conference*, pp. 163–166 (1996).

Chapter 10

Displacement Damage in Compound Semiconductors

Permanent damage at semiconductor/insulator interface regions (ionization damage) is usually the dominant mechanism for silicon devices when they are exposed to radiation. Much of the work on radiation damage is based on tests with cobalt-60 gamma rays, which produce mainly ionization damage. However, most compound semiconductor transistors are only slightly affected by ionization damage because they do not depend on high quality insulators. In many cases a heterojunction buffer layer is used between the surface and the active region that effectively decouples surface recombination from the device, further reducing the effects of ionization damage. Consequently, displacement damage is nearly always the most important damage mechanism for compound semiconductor, in sharp contrast to most silicon devices.

The importance of displacement damage in compound semiconductors is not always recognized. Cobalt-60 gamma rays are often used for radiation testing as well as a reference point for radiation performance in the literature, even for compound semiconductors. Gamma rays can produce changes in these devices, but the damage is usually caused by the relatively small amount of displacement damage produced by gamma rays, not ionization damage.

Such tests will usually underestimate the effects that occur in real space environments. However, SiC MOSFETs are an exception, because they use SiO_2 as a gate insulator, which is sensitive to charge trapping from ionization. just as for silicon MOSFETs.

No standard method exists for radiation tests of compound semiconductors. Radiation tests must be done with high-energy electrons, protons, or neutrons in order to simulate the damage that takes place in space environments. The existing body of work on compound semiconductors uses a plethora of particle types and energies for many reasons, including the availability of specific radiation sources with low cost. This causes a great deal of confusion. A standard reference point is needed in order to compare experimental results in the literature, as well as to apply the results to space environments. We will base our evaluations of device performance on damage from 50-MeV protons, using non-ionizing energy loss (NIEL) to normalize damage from different particle types and energies to the damage produced by that particular proton energy.

Table 10-1 gives approximate factors for normalizing damage in various environments (the fluence of particles in the first column should be multiplied by the factor in the last column to obtain the equivalent 50-MeV proton fluence). Note that the values in the table assume that displacement damage is the dominant mechanism, not ionization damage.

Table 10-1. Factors for Damage Equivalence for Test Data Interpretation (based on GaAs).

Particle Type	Energy (MeV)	Factor for Equivalent Damage from 50-MeV Protons	Fluence Multiplication Factor (Approximate)
Proton	1.8	10.1	0.1
Proton	10	2.03	0.5
Neutron	1 (fission source)	0.25–1*	1–4
Electron	2	1.20×10^{-2}	83
Electron	10	2.48×10^{-2}	40
Cobalt-60 Gamma Ray	1.18 and 1.34 MeV	2.2×10^{-4}	4500

*Experimental results differ widely for neutron damage equivalence.

The values in the table are based on NIEL results for GaAs. The relative damage for other compound semiconductors is very similar, within about 20%, except for wide bandgap devices.

10.1 JFETs

Early radiation studies of III-V transistors were done on GaAs JFETs with uniform channel doping. Damage in those devices could be directly related to carrier removal in the channel region, which caused an increase in threshold voltage and a decrease in transconductance. The experimental results were in good agreement with a model that combined the first-order dependence of JFET characteristics on doping concentration and mobility with simple linear models for carrier removal and mobility degradation.

Later studies on devices with nonuniform channel doping showed a more complicated dependence because carrier removal altered the doping profile [1,2]. A model for threshold voltage was developed, resulting in an exponential dependence between threshold voltage and fluence.

$$V_T = a + be^{c\Phi} \qquad (10\text{-}1)$$

where V_T is threshold voltage, ϕ is particle fluence, and the constants a,b, and c are fitting parameters. An analytical model was also developed that could predict the asymptotic value of V_T at high fluences.

Enhancement-mode JFETs are more difficult to deal with because they use a sharp gaussian channel profile for threshold adjustment. Threshold voltage and transconductance are both affected by radiation damage. Figure 10-1 shows the decrease in transconductance after neutron irradiation [3]. Data points show experimental results, while the lines show calculations based on the effect of carrier removal on the channel profile (the doping levels are average values).

Calculations for a uniform channel predicted degradation at fluences that are about a factor of five lower than those shown in the figure. Relatively high fluences are required in order to produce changes in these devices. The carrier removal rate was modeled using a nonlinear model, where the carrier removal rate depends exponentially on fluence in order to provide better accuracy at high fluence.

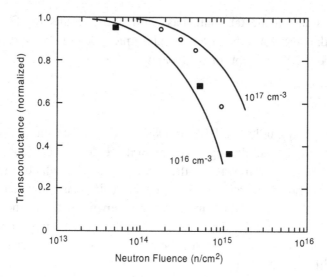

Fig. 10-1. Degradation in transconductance of a GaAs JFET after irradiation with neutrons [1]. © 1978 IEEE. Reprinted with permission.

InP JFETs were evaluated in an earlier study using cobalt-60 gamma rays and 1-MeV electrons [4]. Those devices were evaluated for high-frequency applications, with a decrease of about 5% in small-signal gain at 4.5 GHz at 100 Mrad. There was only a slight difference between the gamma ray and electron results, implying that most of the damage was caused by ionization, not displacement damage.

Note however that the equivalent fluence of 50 MeV protons from a dose of 100 Mrad(InP) of 1-MeV electrons is about 4×10^{13} /cm², which is just below the point where we expect carrier removal to be observable. Thus, the conclusion about ionization damage in that study may be inaccurate. Nevertheless, the older work shows the intrinsic radiation hardness of GaAs and InP JFETs.

10.2 HFETs

More recent tests have been done on GaAs HFETs. They have comparable or higher doping levels compared to JFETs, with similar radiation performance. Only small changes occur until very high fluences are reached, consistent with the carrier removal mechanism.

Figure 10-2 shows the results of neutron tests of AlGaAs HFETs [5]. Very small changes were observed in threshold voltage, even after a fluence of 10^{15} n/cm^2. A model was developed that was in close agreement with the observed changes in threshold voltage, using carrier removal rates of 3 cm^{-1} for the GaAs region and 10 cm^{-1} for the AlGaAs layer. The model assumed that neutron-induced traps were located near the center of the bandgap, and that the carrier removal rate was constant with fluence. Small changes in the depletion width and Fermi level were predicted by the model. Changes in the thickness of the AlGaAs layer were negligible compared to the other factors.

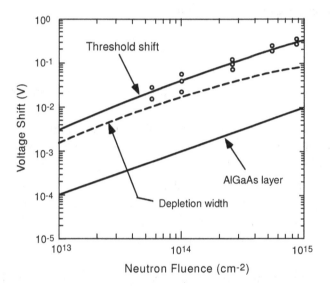

Fig. 10-2. Neutron test results of an AlGaAs/GaAs HFET showing the change in threshold shift and the change in depletion width after neutron irradiation [2]. © 1988 IEEE. Reprinted with permission.

HFETs are complex devices, with many different layers. A more sophisticated model for HFET degradation was developed by Jun and Subramaniam [6], relating device performance to the change in sheet carrier concentration within the device. They used an AlGaAs/GaAs HFET, with an undoped GaAs channel. A triangular potential well was assumed for the 2-DEG region within the device The model only considered the first two bound states. With those assumptions, the sheet charge contribution at the interface, qn_s, is

$$qn_s = \sqrt{q(N_2 - N_{A2})(2\varepsilon_2 v_{20} + qN_2 d_i^2)} - qN_2 d_i - qN_{A1} W \quad (10\text{-}2)$$

where ε_2 is the dielectric constant of AlGaAs, N_2 is the AlGaAs donor concentration, N_{A2} is the AlGaAs acceptor concentration, d_i is the spacer layer, and N_{A1} is the acceptor concentration in the GaAs layer. They found that the main contribution to the sheet charge was the last term in Eq. 10-2; the acceptor layer doping N_{A1} has the lowest doping concentration. Ignoring the square root term, the normalized change in sheet carrier concentration is given by

$$\frac{n_s}{n_{so}} = 1 - \sqrt{\alpha_2 N_{A1}} \quad (10\text{-}3)$$

where α_2 is a degradation constant that is fitted to the measured device degradation.

This result is surprisingly simple, given the complexity of Eq. 10-2, although it is not totally unexpected because of the large differences in carrier concentration for the various materials within the structure.

A comparison of experimental results after neutron degradation is shown in Fig. 10-3, along with predictions using the model of Jun and Subramanium. The vertical axis is the normalized sheet carrier concentration. Carrier removal effects in doped AlGaAs layers and the GaAs buffer region were included in the model (note that carrier removal in a two-dimensional structure is dimensionless, which has to be taken into account when we compare carrier removal in a bulk structure). The maximum fluence (corresponding to the last data point) was 5×10^{14} n/cm^2. The model is in reasonable agreement with experimental results at high fluences, but predicts slightly less damage at low fluences.

They concluded that the dominant mechanism for degradation was a decrease in the acceptor concentration, with a parabolic dependence of carrier removal rate on N_A. They also observed changes in mobility at room temperature that were larger than expected. They postulated that this could be due to carrier scattering from neutral impurities, or from regions with larger order damage.

Both examples of HFET degradation show that the technology is extremely resistant to radiation damage. Models based on carrier removal are sufficiently accurate to predict the onset of degradation at high fluences, but are limited by the uncertainty in the carrier removal coefficient at high fluences. Similar results are expected for HFETs fabricated with other materials. Although carrier removal rates are slightly different for various materials, the high radiation tolerance of HFETs is due to the high doping levels that are used. An example for AlGaN is treated in Section 10.4.

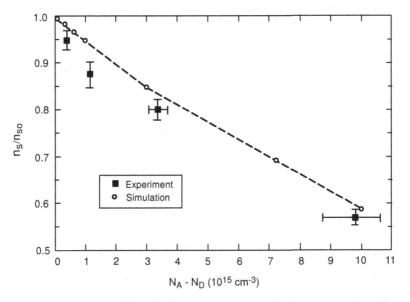

Fig. 10-3. Normalized sheet carrier concentration in the 2-DEG region of an AlGaAs HEMT device vs the difference in acceptor and donor densities in the GaAs buffer region where the 2-DEG gas is formed [6]. © 2002 IEEE. Reprinted with permission.

10.3 Advanced Bipolar Transistors

One of the key applications of advanced bipolar transistors is in high-frequency amplifiers in communication systems. Figure 10-4 shows results for a heterojunction bipolar device – fabricated with silicon-germanium – that was irradiated with protons [7]. This device was designed for RF applications that require operation at high current in order to meet performance goals. Although there is some evidence of degradation at low collector currents, operation at the current range required for optimum RF performance is essentially unaffected up to fluences of 5×10^{13} p/cm^2.

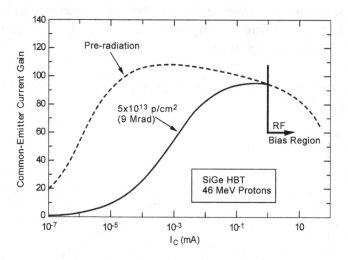

Fig. 10-4. Total dose tests of an advanced HBT with protons [7]. © 1999 IEEE. Reprinted with permission.

Negligible changes were observed in the gain-bandwidth product and standard S-parameters after irradiation, although there were slight increases in noise figure at lower currents. Thus, these devices are highly resistant to radiation damage.

SiGe HBTs are also used for high-speed logic applications. The effect of proton damage on several generations of SiGe HBTs from one manufacturer has been investigated, as shown in Fig. 10-5 [8]. One of the three processes was more sensitive to radiation damage, which was initially unexpected.

The total dose level where changes occur for the 7HP process are lower than the failure level of the 5HP process that preceded it, as well as the more advanced 8HP process. The reason for this was the inclusion of silicon-dioxide material in the lateral isolation region of the 7HP process, which caused gain degradation because of ionization damage in this particular structure. The presence of SiO₂ within this structure makes it an exception to the general rule that compound semiconductor devices are dominated by displacement damage instead of ionization damage.

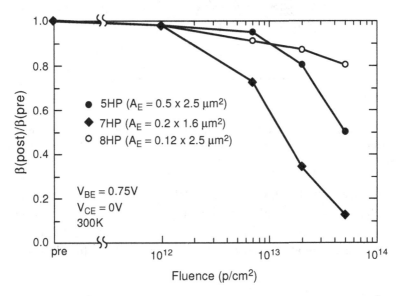

Fig. 10-5. Normalized gain degradation vs. 50-MeV proton fluence for three generations of SiGe HBT devices [8]. © 2003 IEEE. Reprinted with permission.

There are two regions within the 7HP HBT structure where ionization damage can occur, the emitter-base spacer and the edge of the trench isolation region, as shown in the diagram of Fig. 10-6 [9]. Charge trapping in the emitter-base spacer oxide was determined to be the mechanism for increased sensitivity for the 7HP process. The subsequent 8HP process used a different geometry, eliminating the spacer oxide. This increased the radiation hardness, making it comparable to that of the earlier 5HP process from the same manufacturer. Note that these are commercial devices, not special devices made for space applications. The geometrical changes were made to improve electrical performance.

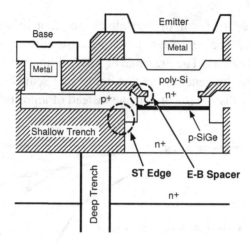

Fig. 10-6. Diagram of the HBT in the 7HP process showing the two regions where ionization damage can affect transistor gain [9]. © 2002 IEEE. Reprinted with permission.

Advanced bipolar devices can be fabricated with other materials as well. Radiation tests of InP heterojunction bipolar transistors have shown similar results. Figure 10-7 shows that very little degradation occurs in such devices, even after exposures to levels above 100 Mrad(InP) [10].

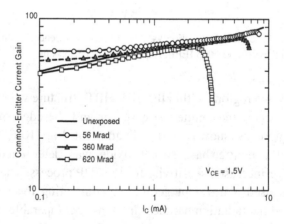

Fig. 10-7. Total dose degradation of an advanced InP bipolar transistor [10]. Although the results are given in rads, the equivalent displacement damage for the highest fluence is within the range where carrier removal becomes important. © 1999 IEEE. Reprinted with permission.

These experiments were done using high-energy electrons, which introduce about 40 times less damage than 50-MeV protons. A fluence of 56 Mrad(Si) corresponds to an equivalent proton fluence of about 2×10^{13} (50 MeV) p/cm^2. Thus, the highest total dose level in this work overlaps the region where we expect carrier removal to be a factor. These devices are very resistant to radiation damage, and would only be affected at extremely high radiation levels.

10.4 Wide-Bandgap Devices

A limited amount of work has been done on radiation effects in wide bandgap devices. Various technologies have been investigated, including JFETs, MOSFETs (restricted to SiC because of material considerations), and GaN HFETs. The devices that have been studied represent experimental results for evolving technologies, not production devices.

10.4.1 Silicon carbide

10.4.1.1 SiC MOSFETs

Figure 10-8 shows the degradation of a SiC JFET after it was irradiated with neutrons. Very high radiation levels are required before measurable damage occurs [11]. The main effect was an increase in channel resistance, not changes in threshold voltage. The carrier removal rate that was measured for tests done at room temperature was 3.5 cm^{-1}, about an order of magnitude lower than the carrier removal rate for GaAs JFETs. Analysis of these results is complicated because the donors are not fully ionized at room temperature. The devices were fabricated with 6H SiC. Changes in the Fermi level that are caused by defects from radiation also affect the number of ionized donors.

SiC MOSFETs have also been evaluated, using cobalt-60 gamma rays, which is appropriate for those structures because they use SiO_2 as a gate insulator. Figure 10-9 shows how the gate threshold voltages of p- and n-channel SiC MOSFETs are affected by ionization damage [12]. The gates were formed using pyrogenic oxides, with a gate oxide thickness of 33 nm. Initial gate leakage currents were 0.1 pA or less.

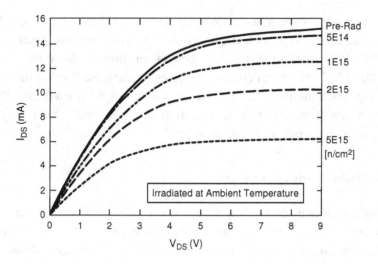

Fig. 10-8. Neutron damage of SiC JFETs [11]. © 1994 IEEE. Reprinted with permission.

Much larger changes in threshold voltage occurred for p-channel devices because of the negatively charge interface states that are created by the radiation. The interface states partially compensate the oxide trapped charge for NMOS devices, which is the reason for reduced radiation sensitivity for that structure. Both types of MOSFETs have smaller radiation-induced changes than conventional silicon MOSFETs. However, the radiation tests were done with the gates floating (unconnected), an unusual condition that will likely reduce hole trapping compared to the case where the gate is biased in a manner that enhances charge transport towards the SiC/SiO_2 interface region.

SiC MOSFETs are continuing to evolve, and are now commercially available. They are mainly intended for applications with high voltage and high current, which require thicker gate regions compared to mainstream silicon logic devices.

Devices with thicker oxides will likely be more affected by total dose damage compared to the results shown in Fig. 10-9. As noted previously, bias conditions on the gate during and after irradiation affect the magnitude of the change in gate threshold voltage, an important consideration.

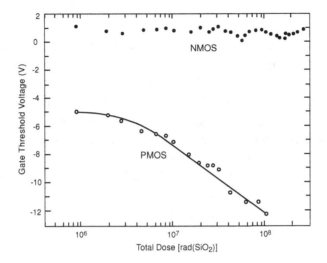

Fig. 10-9. Threshold voltage vs. total dose for SiC MOSFETs with pyrogenic gate oxides [12]. © 2003 IEEE. Reprinted with permission.

10.4.1.2 Schottky barrier diodes

Displacement damage has also been studied in SiC Schottky barrier diodes, which are commercially available [13]. Devices with voltage ratings from 300 to 1200 V were included, all with a forward current rating of 10 A. The devices were fabricated with 4H silicon carbide, and irradiated with 205 MeV protons.

The results showed that the turn-on voltage and reverse characteristics were essentially unchanged by radiation, but that large increases occurred in the series resistance at high radiation levels. The devices with the highest voltage rating (1200 V) were affected at lower radiation levels than the two devices with lower voltage ratings, consistent with general expectations for the effect of carrier removal on resistance. The series resistance of the 1200 V devices increased by a factor of 10 at a fluence of approximately 2.5×10^{14} p/cm^2. The changes became strongly nonlinear at high fluences, particularly for the devices with the high voltage rating. However, this is a very high fluence, and we can conclude that high-voltage SiC Schottky diodes are far more resistant to displacement damage than similar devices fabricated with silicon.

10.4.2 Gallium nitride

Proton damage and annealing in AlGaN HFETs were investigated by Cai, *et al.* [14]. The devices were irradiated to a fluence of 10^{14} p/cm^2 with 1.8 MeV protons (protons at that energy are more than 20 times more damaging than 50-MeV protons, and this is a very high particle fluence).

As shown by the initial change at room temperature in Fig. 10-10, the very high radiation level reduced the saturation current to about 30% of its pre-irradiation value. The damage was stable at room temperature. Temperatures above 300°C were required in order to anneal some of the damage, and the annealing rate was low until the temperature was raised to 600°C. They suggested that lattice strain may play a role in annealing at very high temperature.

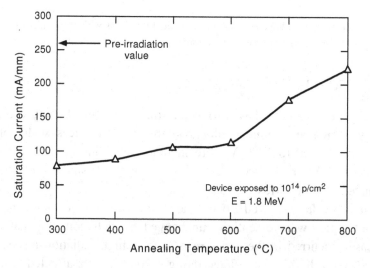

Fig. 10-10. Annealing of proton damage in an AlGaN HFET [14]. © 2000 IEEE. Reprinted with permission.

A later study of AlGaN HFETs was done by Hu, *et al.*, also using 1.8 MeV protons [15]. The devices that they tested used a thin AlN layer between the semi-insulating GaN layer and the AlGaN layer that reduced the spontaneous polarization and the resulting discontinuity in the conduction band. Figure 10-11 shows the effect of high levels of proton radiation on transconductance and threshold voltage for those devices.

From the standpoint of device performance, the main advantage of the additional AlN layer is a 60% increase in sheet carrier mobility. Hu, et al. [15], were able to model the effect of radiation on mobility using a linear relationship between sheet carrier mobility and threshold voltage. They postulated that defect centers near the 2-DEG gas region may scatter carriers within the 2-DEG because of Coulomb interactions, reducing the carrier mobility, as well as reducing sheet carrier mobility.

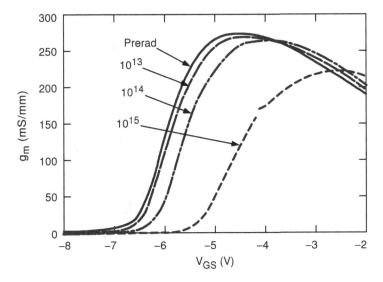

Fig. 10-11. The effect of 1.8 MeV protons on transconductance for AlGaN HFETs [15]. © 2003 IEEE. Reprinted with permission.

A recent study considered the interplay between displacement damage and the polarization charge in AlGaN HFETs with an AlN layer [16]. That work showed that electrostatic effects between deep acceptor-like defects in the AlGaN layer were the main cause of degradation of the 2DEG properties of the devices, an interesting theoretical result.

The effect of gamma irradiation on GaN Schottky diodes has also been evaluated [17]. Figure 10-12 shows the increase in reverse leakage current that occurred after irradiation in that study. The forward characteristics were unchanged. Deep-level transient spectroscopy analysis showed that the gamma rays introduced shallow defects with an introduction rate of 2.12 x 10^{-3} cm^{-1}, consistent with carrier removal from displacement effects when we take the very low NIEL value of gamma rays into account. The reverse current annealed after about one week at room temperature, along with a 15 minute interval at 50°C.

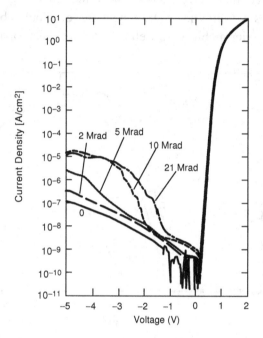

Fig. 10-12. Effects of high levels of gamma radiation on GaN Schottky diodes [17]. © 2003 IEEE. Reprinted with permission.

The C-V characteristics were changed by radiation, which was attributed to an increase in the Schottky barrier height, but no significant change was observed in the forward characteristics. They suggested that the near constant Schottky barrier height in the forward direction was caused by the large number of dislocations that are present in the material, which are terminated at surfaces.

These results show that Schottky interfaces in GaN technology are somewhat more sensitive to radiation damage than the underling semiconductor layers. This could be an important limitation for applications that require operation at high radiation levels.

10.5 Summary

This chapter has discussed displacement damage effects in compound semiconductor transistors. With the exception of SiGe HBTs and SiC MOSFETs, which are sensitive to ionization damage as well as displacement damage, carrier removal is the dominant mechanism for device degradation. Although several different types of radiation have been used to evaluate radiation damage by various studies, radiation levels above the displacement damage equivalent of about 10^{13} 50-MeV protons per cm^2 are generally required before measurable damage occurs in compound semiconductors, a very high radiation level.

This result is consistent with the doping levels that are generally used in the critical active layers, along with the carrier removal mechanism and the insensitivity of most compound semiconductors to radiation-induced surface traps.

Advanced modeling methods have been developed for degradation in HFETs that relate the sheet carrier concentration to bulk carrier removal rate, providing reasonable agreement between theory and experimental results.

A simple linear coefficient for the carrier removal rate appears to be adequate for developing models that show where degradation first starts to take place in JFETs and HFETs. However, that assumption only holds if the change in carrier density is approximately 25% or less, and it is necessary to use nonlinear models for accurate modeling of the region in which substantial damage takes place.

Unlike the other technologies, SiGe HBTs are affected by ionization damage. One example shows a particular design variation that caused significant radiation damage at total dose levels below 1 Mrad(Si). However, the majority of SiGe HBTs will withstand total dose levels >1 Mrad(Si). Ionization damage is also important for SiC MOSFETs, which use SiO_2 gate oxides.

Despite the various ways in which radiation tests have been done in the literature, all of the results verify that compound semiconductors are very resistant to permanent damage from radiation. Transient radiation effects, which are far more important for these devices, will be discussed in Chapter 13.

References

1. R. Zuleeg, J. K. Notthoff and K. Lehovec, "Radiation Effects in Enhancement Mode GaAs Junction Field Effect Transistors", IEEE Trans. Nucl. Sci., **24**(6), pp. 2305–2308 (1977).
2. A. B. Campbell, *et al.*, "Particle Damage in GaAs JFET Test Structures", IEEE Trans. Nucl. Sci., **33**(6), pp. 1435–1441 (1986).
3. R. Zuleeg and K. Lehovec, "Neutron Degradation of Ion-Implanted and Uniformly Doped Enhancement Mode GaAs JFETs", IEEE Trans. Nucl. Sci., **25**(6), pp. 1444–1449 (1978).
4. W. T. Anderson and J. B. Boos, "Radiation Effects in InP JFETs", IEEE Trans. Nucl. Sci., **32**(6), pp. 4001–4004 (1985).
5. B. J. Krantz, W. L. Bloss and M. J. O'Loughlin, "High Energy Neutron Effects in GaAs Modulation-Doped Field Effect Transistors (MODFETs)", IEEE Trans. Nucl. Sci., **35**(6), pp. 1438–1443 (1988).
6. B. Jun and S. Subramanium, "Carrier Removal Rate and Mobility Degradation in Heterojunction Field-Effect Transistor Structures", IEEE Trans. Nucl. Sci., **49**(6), pp. 3222–3229 (2002).
7. S. Zhang, *et al.*, "The Effects of Proton Irradiation on the RF Performance of SiGe HBTs", IEEE Trans. Nucl. Sci., **46**(6), pp. 1716–1722 (1999).
8. Y. Lu, *et al.*, "Proton Tolerance of Third-Generation 0.12 μm 185 GHz SiGe HBTs", IEEE Trans. Nucl. Sci., **50**(6), pp. 1811–1815 (2003).
9. J. D. Cressler, *et al.*, "An Investigation of the Origins of the Variable Proton Tolerance in Multiple SiGe HBT BiCMOS Technology Generations", IEEE Trans. Nucl. Sci., **49**(6), pp. 3203–3207 (2002).
10. A. Shatalov, *et al.*, "Electron Irradiation Effects in Polyimide Passivated InP/InGaAs Single Heterojunction Bipolar Transistors", IEEE Trans. Nucl. Sci., **46**(6), pp. 1707–1714 (1999).
11. F. B. McLean, *et al.*, "Analysis of Neutron Damage in High Temperature Silicon Carbide JFETS", IEEE Trans. Nucl. Sci., **41**(6), pp. 1884–1890 (1994).
12. K. K. Lee, T. Ohshima and H. Itoh, "Performance of Gamma Irradiated P-Channel 6H-SiC MOSFETs: High Dose", IEEE Trans. Nucl. Sci., **50**(1), pp. 194–200 (2003).
13. R. D. Harris, A. J. Frasca and M. O. Patton, "Displacement Damage Effects on the Forward Bias Characteristics of SiC Schottky Barrier Power Diodes", IEEE Trans. Nucl. Sci., **52**(6), pp. 2404–2412 (2005).
14. J. S. Cai, *et al.*, "Annealing Behavior of a Proton Irradiated $Al_xGa_{1-x}N$/GaN High Electron Mobility Transistor Grown by MBE", IEEE Trans. Elect. Dev., **47**(2), pp. 304–306 (2000).

15. X. Hu, *et al.*, "Proton-Irradiation Effects on AlGaN/AlN/GaN High Electron Mobility Transistors", IEEE Trans. Nucl. Sci., **50**(6), pp. 1791–1796 (2003).
16. A. Kalavagunta, *et al.*, "Electrostatic Mechanisms Responsible for Device Degradation in Proton Irradiated AlGaN/AlN/GaN HEMTs", IEEE Trans. Nucl. Sci., **55**(4), pp. 2106–2112 (2008).
17. G. A. Umana-Membreno, *et al.*, "^{60}Co Gamma Irradiation Effects on n-GaN Schottky Diodes", IEEE Trans. Elect. Dev., **50**(12), pp. 2326–2334 (2003).

Chapter 11

Displacement Damage in Optoelectronic Devices

Displacement damage in optoelectronic devices varies over an extremely wide range. Some types of devices are among the most sensitive to displacement damage effects, degrading significantly even in low-earth orbits, where radiation levels are relatively low. Others are extremely tolerant to radiation damage. This wide range in radiation damage sensitivity is related to device design and basic operating principles [1–5]. Although it is relatively straightforward to discuss the mechanisms involved, often the specifications of optoelectronic devices include little or no information about fabrication methods or device construction, making it inherently more difficult to deal with displacement damage in optoelectronics compared to conventional electronic devices. Radiation testing is often required to verify satisfactory performance for optoelectronic devices in typical space environments because of the limited information that is usually available about device design.

Special diagnostic methods can often be used to supplement radiation testing, as well as to identify devices that are potentially more sensitive to radiation damage. Several diagnostic techniques will be discussed in this chapter, along with the mechanisms that affect radiation damage.

As noted in Chapter 9, calculations of non-ionizing energy loss (NIEL) do not agree with experimental results for several types of optoelectronic devices [6]. The discrepancy is very large for protons with energies >50 MeV, as well as for neutrons. We will not discuss this further, but caution the reader about the importance of these discrepancies for these particular classes of devices. Nearly all of the experimental results in this chapter are for 50 MeV protons, where the NIEL discrepancy is of little importance.

11.1 Light-Emitting Diodes with Amphoteric Doping

11.1.1 Properties affected by radiation damage

Amphoterically doped LEDs are more affected by radiation than nearly any conventional electronic component. That extreme sensitivity is the result of the fabrication method (see Section 4.3.2). The key point is that these types of LEDs have a broad transition region between the n- and p-type materials, with a typical junction width of about 50 µm. They require long lifetimes in order to operate efficiently, and consequently are highly sensitive to lifetime damage.

Radiation damage increases the number of non-radiative recombination centers. This decreases the optical power output at a given current level, because some of the forward current is diverted into non-radiative rather than radiative recombination. For an ideal homojunction diode we can use a simple model that divides the forward current density into two components, one associated with diffusion current, which produces optical power, and the other associated with non-radiative recombination current

$$J = k_1 \exp(qV_F / kT) + k_2 \exp(qV_F / 2kT) \qquad (11\text{-}1)$$

where k_1 and k_2 are the coefficients of the two terms.

The factor of two difference in the exponential factors of these terms causes the forward voltage to decrease at a specific current value when the recombination term (with coefficient k_2) increases. Thus, we expect the decrease in optical output power after degradation to be accompanied by a small decrease in forward voltage when measurements are made at a fixed current. We shall return to this later, pointing out its use as a diagnostic method.

The most important effect of radiation damage is reduction in optical power. Figure 11-1 shows an example for an 880 nm AlGaAs LED, after exposure to 50-MeV protons. The devices were unbiased during irradiation to reduce injection-enhanced annealing, which would otherwise cause a significant fraction of the damage to recover. Measurements were made before and after each irradiation at various forward currents, using pulsed measurement techniques with a low duty cycle. The maximum allowable steady state current of this device is 100 mA, which is usually derated in order to maintain high reliability. Thus, operation at lower currents is a key concern for space applications.

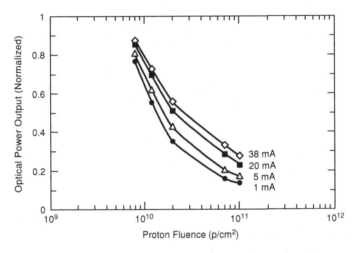

Fig. 11-1. Degradation of an AlGaAs LED (880 nm) after irradiation with 50-MeV protons. The devices were unbiased during irradiation, and measured at the forward currents shown.

Less degradation occurs when the device is operated at higher forward current. However, only about ½ the optical power remains after irradiation to levels of 1–2×10^{10} p/cm^2 even in applications with higher forward current.

Often there is considerable variability in the damage of different devices from a specific production lot. An example is shown in Fig. 11-2 for an AlGaAs LED, measured at a forward current of 1 mA, and irradiated with 50 MeV protons. The solid line shows the mean fractional degradation of optical power from a group of 84 devices, while the dashed lines show the best and worst devices from the group.

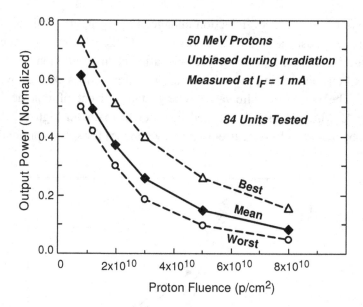

Fig. 11-2. Variability in damage of amphoterically doped AlGaAs LEDs after irradiation with 50 MeV protons.

The unit-to-unit variability appears to be caused by differences in minority carrier lifetime. There is a strong correlation between the initial lifetime before irradiation and the amount of damage that occurs; devices with longer lifetime are more susceptible.

Measurements of reverse-recovery time can be used to identify devices that are more sensitive to radiation damage, providing an effective electrical screening method, as shown in Fig. 11-3 [7]. Typical initial lifetime values for amphoterically doped AlGaAs or GaAs LEDs are in the range of 0.8 to 2 μs. The lifetime of heterojunction LEDs is much shorter, ranging from approximately 5 ns to 0.2 μs, depending on the specific technology.

11.1.2 Damage linearity

As discussed in Chapter 8, the change in reciprocal lifetime in a semiconductors is usually linear with particle fluence, which was used to define the lifetime damage constant (Section 9.2.1). That same relationship also applies to LEDs, as shown in Fig. 11-4. Reverse recovery time measurements were used to measure the lifetime.

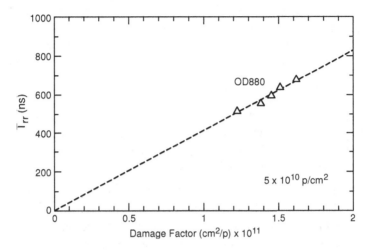

Fig. 11-3. Correlation between reverse recovery time and damage factor for AlGaAs LEDs (880 nm) [7].

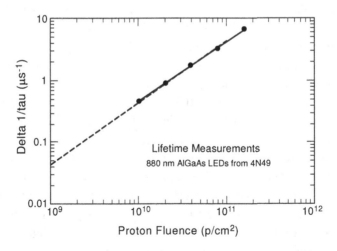

Fig. 11-4. Degradation of minority carrier lifetime of an amphoterically doped LED, using reverse recovery time measurements.

Even though lifetime damage has the expected linear dependence on fluence, the effects of radiation damage are less straightforward when we consider optical power. Rose and Barnes showed that under conditions of constant injection (constant forward current during measurements), LED output power degradation could be related to lifetime damage by the equation [1]

$$\left(\frac{I_o}{I}\right)^n - 1 = \tau_o K \Phi \qquad (11\text{-}2)$$

where I_o is the initial power output before irradiation, I is the reduced light output after irradiation, τ_o is the pre-irradiation value of minority carrier lifetime, K is the damage constant, Φ is particle fluence, and n is an exponent that depends on the properties of the LED.

For the case where optical power and forward current both depend on diffusion of minority carriers, they showed that n = 2/3, provided that the radiation-induced defects are uniformly distributed throughout the forbidden region of the bandgap. The implication of n < 1 is that the optical power degradation is *superlinear* with fluence, increasing as $\Phi^{(1.5)}$ when n = 2/3. Thus, even though the lifetime degrades linearly with fluence (see Fig. 11-4), we expect a superlinear dependence of optical power degradation on particle fluence. This applies to any LED where the efficiency is limited by minority carrier diffusion.

An example of the application of Eq. 11-2 to an AlGaAs LED (880 nm wavelength) is shown in Fig. 11-5. With n = 2/3 the slope is almost exactly one, verifying the applicability of Eq. 11-2. As an aid in interpreting these results, when the value of the ordinate is 1 the optical power has decreased to 63% of the initial value; when the ordinate is 10, the optical power is only about 20% of the starting value.

Rose and Barnes did not explicitly discuss heterostructure LEDs. However, the discussion of recombination current in their study implies that devices with thin regions will have a different exponent, approximately one, because diffusion is no longer the dominant process for forward current injection. Thus, we expect heterostructure LEDs to have a different dependence on fluence compared to LEDs with thick transition regions and long lifetime.

Although using Eq. 11-2 is more complicated compared to the normalized power output used to describe damage in Figs. 11-1 and 11-2, the linear relationship with fluence provides a quantitative way to compare damage of different devices. It is particularly important for amphoterically doped LEDs because we can use the equation to define a damage factor that is applicable over a wide range of fluences. Departures from linearity take place at very high fluences, due to other mechanisms, such as changes in the depletion region width because of carrier removal, but the linear relationship applies over most of the useful range of the LED.

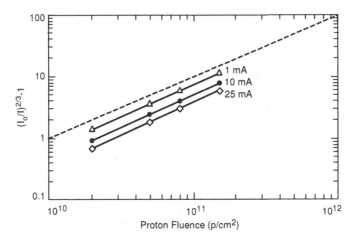

Fig. 11-5. Degradation of an amphoterically doped AlGaAs LED using Eq. 11-2 to linearize the damage.

11.1.3 Annealing

Only very slight recovery (annealing) takes place in light-emitting diodes as long as they are not actively biased, even after several months of storage under unbiased conditions at room temperature. However, a substantial fraction of the damage in amphoterically doped LEDs will recover when forward current is applied after irradiation. Less time is required at high currents compared to recovery at low current; to first order the recovery depends on the total charge that flows through the device after irradiation. Figure 11-6 shows an example, where currents between 5 and 50 mA were applied to an AlGaAs LED (880 nm) with a maximum forward current rating of 100 mA [8].

The proton fluence was high enough to reduce the light output to about 9% of its initial value when the irradiation was stopped. Power output was then measured vs. time for different samples, using various forward currents. The light output increased to approximately 17% of initial value after a long period under forward bias, roughly doubling the post-irradiation light output.

For these devices, a total charge of about 8 C was required in order for ½ of the recoverable damage to occur. Similar results were obtained for GaAs LEDs with a wavelength of 930 nm.

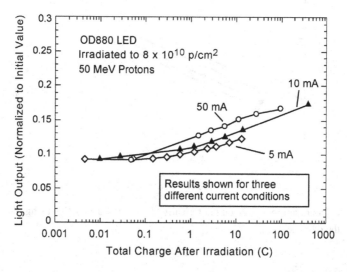

Fig. 11-6. Effect of current injection after irradiation on light output of an amphoterically doped AlGaAs LED with a nominal wavelength of 880 nm. The results at different current can be normalized to the total injected charge [8].

Different annealing properties are observed for damage from various radiation sources because the microscopic defect structure is different. Figure 11-7 compares annealing of amphoterically doped LEDs that were irradiated using different types of radiation [9]. The fluence was selected to reduce the light output to about 20% of the initial value.

Cobalt-60 gamma rays which produce displacements through Compton electrons, produce Frenkel defects. Those defects not only anneal faster, but a much larger fraction of the damage recovers compared to annealing from more energetic particles that produce a mix of Frenkel defects and cascade damage regions.

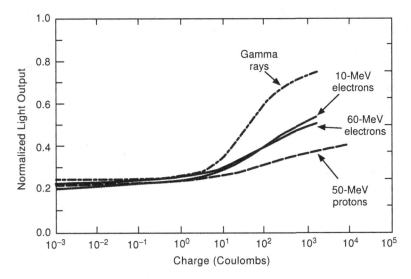

Fig. 11-7. Comparison of annealing in an amphoterically doped LEDs irradiated with different types of particles [9].

Injection-enhanced annealing can be used to reduce damage in space. For example, an LED failed after more than 15 years of operation in the Galileo space probe (which operated in the intense radiation belts of Jupiter). This was corrected by modifying the operating mode to forward bias the device for an extended period, allowing sufficient recovery in the damaged device to restore normal operation [10].

Although annealing can be beneficial in selected applications, the strong recovery of radiation damage during high injection can affect experimental results, and needs to be taken into account during radiation testing. Tests that are done using biased devices will underestimate the amount of damage that will occur for unbiased devices, or for parts that are used at lower forward currents or under pulsed conditions.

11.2 Heterojunction LEDs

LEDs fabricated with heterojunctions are far more resistant to radiation damage because they have shallow active regions, enabling them to operate with shorter minority carrier lifetimes. This is partly due to the band offset of the different material types. As discussed in Chapter 4, they are less efficient than amphoterically doped LEDs, requiring higher current densities to produce measurable light output. Often the first evidence of radiation damage is provided by the decrease of forward voltage at low injection, not degradation of light output. This is illustrated for an 850 nm AlGaAs LED in Fig. 11-8.

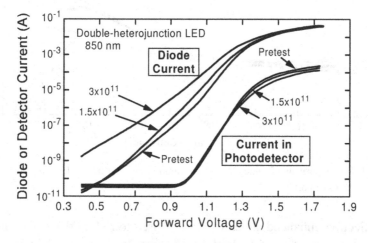

Fig. 11-8. Degradation of a double heterojunction LED showing the decrease in forward voltage at low injection. The output power was measured with an external photodetector.

Figure 11-9 compares typical results for three different heterojunction LEDs with degradation of an amphoterically doped LED [7]. All four LEDs have wavelengths within the range of conventional silicon detectors, and are somewhat interchangeable for many applications. The heterojunction LEDs can withstand radiation levels that are 20–100 times higher. All of the LEDs are fabricated with AlGaAs.

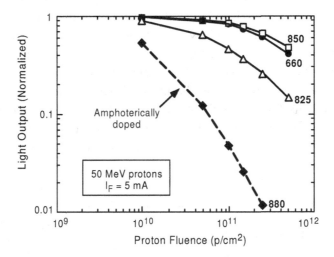

Fig. 11-9. Proton degradation of several double-heterojunction AlGaAs LEDs compared with degradation of an amphoterically doped LED [8].

The improved radiation performance of heterojunction LEDs would appear to be a convincing argument against using amphoterically doped LEDs. However, the high efficiency of amphoterically doped LEDs provides up to a factor of ten more light output than the other devices prior to irradiation. Thus, there is a tradeoff between the high initial efficiency of amphoterically doped devices and their high sensitivity to radiation damage. They can still be used in moderate radiation environments, providing the larger relative degradation can be accommodated in the application.

Equation 11-2 can also be applied to the heterojunction LEDs in Fig. 11-9. The best linear fit for heterojunctions occurs for exponents between 0.9 and 1, supporting the conclusion that lifetime damage is not the controlling damage mechanism for those devices.

Annealing characteristics are also different. Double-heterojunction LEDs are relatively insensitive to annealing. Although annealing has not been studied as extensively for double-heterojunction LEDs, typically less than 10% of the damage will recover from forward injection, even after very long time periods.

Limited evaluations have been done on LEDs operating at longer wavelengths. Figure 11-10 shows results for an InGaAsP LED with a wavelength of 1550 nm [11]. The radiation sensitivity of that device is comparable to that of the 850 nm and 660 nm devices shown in Fig. 11-8. However, damage in the InGaAsP LED is nearly linear with n = 2/3, just as for the amphoterically doped LEDs discussed earlier, event though the InGaAsP LEDs are about two orders of magnitude less sensitive to radiation damage. This implies that lifetime damage is the dominant mechanism for these structures.

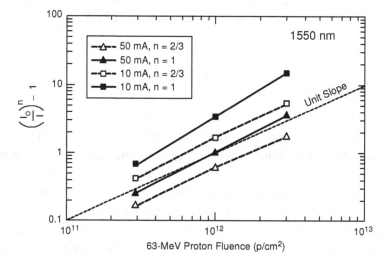

Fig. 11-10. Degradation of an AlGaAs LED using Eq. 11-2 to linearize the damage [11].

Recent work has also been done on GaN LEDs, operating at 430 nm [12]. Normalized light intensity vs. fluence using 10-MeV protons is shown for those devices in Fig. 11-11, along with data for bulk GaAs (amphoterically doped) and quantum-well GaAs LEDs. The GaN devices are much more resistant to radiation damage, requiring a fluence that is about twenty times higher than that of heterojunction GaAs LEDs to produce the same damage. Damage constants for GaN are expected to be much higher than for GaAs because of the higher displacement energy threshold for that material, but have not been studied in detail.

It is somewhat unusual to use 10-MeV protons, partly because they have very limited range. Fortunately, the study compared GaAs and GaN devices at the same proton energy, and there is no ambiguity about the higher damage constant of GaN. If we assume that the dependence of NIEL on energy is the same for GaN and GaAs, then the fluence values in Fig. 11-11 should be multiplied by 4.7 to get the equivalent fluence for 50-MeV protons.

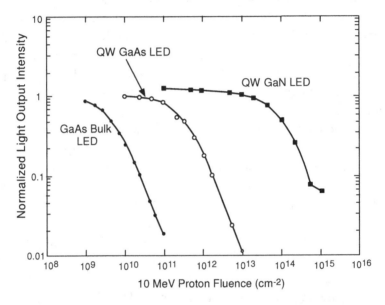

Fig. 11-11. Experimental results for GaN LEDs, emitting at 430 nm. The other two curves are for GaAs LEDs, with wavelengths of approximately 850 nm [12].

Finally, it should be noted that LEDs degrade in a graceful way after they are damaged from radiation. They are usually used at constant current. Under that condition the light output degrades, along with small changes in the forward voltage. Power dissipation actually decreases in damaged devices, and there is no reason to expect that even severe degradation in light output will affect reliability. That is not the case for laser diodes, as discussed in the next section.

11.3 Edge-Emitting Laser Diodes

11.3.1 Fundamental effects

Laser diodes operate at much higher electrical and optical power densities compared to LEDs. Their operation also involves different fundamental parameters, particularly the dependence of cavity gain on current density, which is related to fundamental material properties. A laser cannot function unless the internal photon density is high enough to reach the point where stimulated emission takes place. It is clear from the discussion of radiation effects on LEDs that even higher internal carrier densities will be needed when radiation-induced recombination centers divert part of the current into non-radiative processes. Thus, we expect that the threshold current of a laser will increase after it is damaged by radiation.

Threshold current is a key parameter for laser diodes, along with the linearity (slope efficiency) of the optical power output above the threshold current level. Unlike LEDs, which can tolerate relatively large changes in output power, much smaller changes are permissible for laser diodes because the power density is so much higher.

Figure 11-12 shows how proton degradation affects the optical power characteristics of a laser diode [13]. The threshold current begins to change after the device is irradiated to a fluence of about 3×10^{12} p/cm^2. There are slight changes in slope efficiency as well. Note also that there is some "softening" of the transition from LED to laser operation after irradiation.

Laser diode degradation can be analyzed by considering the recombination rate just below the threshold condition

$$R = An + Bn^2 + Cn^3 = \frac{n_{th}}{\tau} \qquad (11\text{-}3)$$

where A is the bulk recombination rate (non-radiative), B is the spontaneous band-to-band recombination rate, C is the Auger recombination rate, n is the carrier concentration, n_{th} is the carrier concentration at threshold, and τ is the effective lifetime.

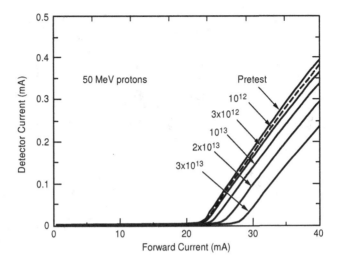

Fig. 11-12. Effect of proton irradiation on optical power output of a 650 nm laser diode [13].

The coefficient B has approximately the same value, $\sim 10^{-10}$ cm^3/s, for AlGaAs, InGaAs and InGaAsP. As a result, carrier densities above 10^{18} cm^{-3} are required to increase the spontaneous recombination rate to the point where the photon density is sufficiently high to reach threshold for all three material systems. When the bulk recombination rate increases due to radiation damage, the carrier concentration must increase to overcome the increase in the loss term A (that term is linear with n in Eq. 11-3). Thus, the carrier density is higher after radiation damage, affecting all terms in the equation. For materials with high Auger recombination, there is effectively a race between the n^2 dependence of spontaneous radiative recombination and the n^3 dependence for Auger losses as n increases. Eventually the Auger term will dominate, limiting the light output. The Auger loss term is effectively an upper limit on carrier density. With extreme degradation, a laser will effectively stop operating if the non-radiative loss term is sufficiently high.

Semi-logarithmic plots of laser diode characteristics show that the first-order effect of radiation damage is to increase the non-radiative recombination term, as shown in Fig. 11-13 [13].

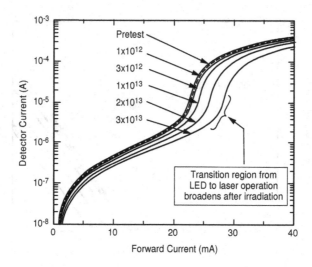

Fig. 11-13. Results of Fig. 11-12 plotted with a logarithmic vertical axis.

Below the threshold region (where the laser functions as an LED), light output is reduced because of the increased number of N.R. recombination centers within the active region after irradiation. More current is required to overcome those loss terms, but the optical power level required to initiate laser operation is nearly unchanged. Threshold current increases in a nearly linear way with radiation damage as long as the changes are small enough so that second-order terms in the exponential relationship can be ignored. However, the relationship becomes nonlinear when threshold current changes approach 40%.

Laser diodes are also affected by current-enhanced annealing [14]. This is somewhat surprising because double-heterojunction LEDs, fabricated with similar doping levels and dimensions, are relatively insensitive to annealing. Figure 11-14 shows annealing for a 1300 nm laser diode, after irradiation with protons which increased the threshold current by about 30%. Annealing proceeds more rapidly when the current injected during annealing is above the threshold current (solid symbols), compared to a current of 5 mA, which is below the threshold for laser operation. This suggests that the very high optical power density that is present in laser diodes also affects annealing properties, along with the injection level.

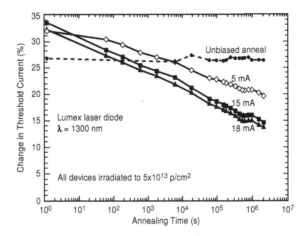

Fig. 11-14. Dependence of laser diode threshold current on time for various injection conditions.

The change in threshold current is nearly linear with fluence for most laser diodes. This linearity extends up to the point where the threshold current has increased by about a factor of three, which is severe enough to have a large effect on device reliability in typical applications.

Figure 11-15 shows the threshold current dependence on fluence for four different laser diodes, using different materials [13]. The fluence for 10% increase is nearly the same for the three in-plane devices, despite the different materials used in their construction. One explanation for this is that the threshold carrier density for laser operation is (coincidentally) nearly the same for the three materials in this figure. Carrier removal rates are also of the same order. However, one would expect different results for materials such as GaN where the damage constant is lower and threshold conditions depend on polarization charge as well as carrier density.

Additional diagnostic information can be obtained by examining the derivative of optical power with respect to current. Very precise measurements are required in order to avoid scatter in the derivative, but derivative plots show several key properties, including the threshold current, the transition behavior from LED to laser mode, and the slope efficiency.

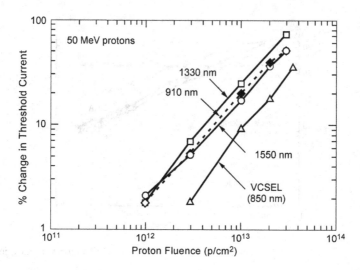

Fig. 11-15. Threshold current degradation of several different laser technologies [13].

Figure 11-16 shows an example of a derivative plot before and after irradiation for a 1500 nm laser diode. The device was irradiated with 50-MeV protons.

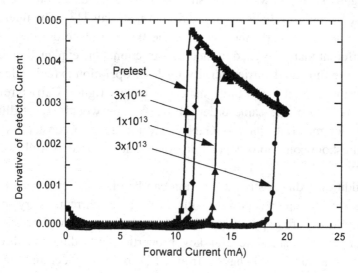

Fig. 11-16. Derivative plot for a 1500 nm laser before and after irradiation.

Although threshold current is the most important parameter, slope efficiency is also affected by radiation damage. In most cases changes in that parameter are relatively small, as illustrated in Fig. 11-17 for 1300 nm and 1550 nm laser diodes. Fluences above 3×10^{13} p/cm^2 are required in order to decrease slope efficiency by more than 5%, which is approximate fluence range where threshold current doubles for typical laser diode technologies.

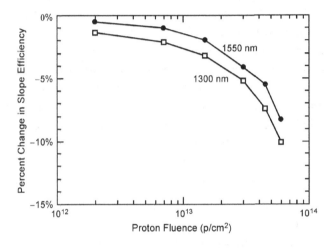

Fig. 11-17. Decrease in slope efficiency for 1300 and 1550 nm laser diodes after irradiation with 50-MeV protons.

Another important parameter for laser diodes is temperature sensitivity. InGaAsP used in 1300 and 1550 nm lasers has a higher Auger coefficient, and increased temperature sensitivity compared to AlGaAs. This leads to the expectation that the threshold current will be more sensitive to temperature for damaged devices. However, experimental results showed that temperature sensitivity actually decreased after irradiation [13]. One reason for this is that the fraction of the current near threshold that is diverted into non-radiative recombination is effectively an offset term, and as it increases the apparent temperature sensitivity will decrease because we are now normalizing the threshold current to a larger "base" term when the temperature sensitivity is measured after irradiation. Unless the laser is pushed to the point where the Auger term becomes important, little change is expected in the temperature sensitivity from radiation damage. InGaAsP lasers are typically designed to operate well below that region.

11.3.2 Monitor diode and operational margins

Laser diodes often incorporate internal monitor diodes that provide a method of monitoring the optical power output. The monitor diode signal can be used as part of a control system to stabilize the laser output because of the effects of temperature, aging, or other factors. However, in many cases the internal monitor diode is more sensitive to radiation damage than the laser. Figure 11-18 shows an example for a 1300 nm laser diode that has a compound semiconductor diode monitor. Degradation of the monitor diode could adversely affect the reliability of the laser diode unless the degradation of the monitor diode is taken into account in the system application.

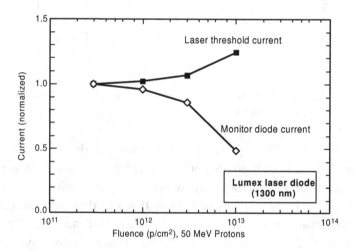

Fig. 11-18. Degradation of the internal monitor diode of a 1300 nm laser diode.

Monitor diodes are often discrete components which are attached to the laser structure after initial mounting and testing. Thus, the monitor diode can be fabricated from different materials compared to those required for the laser diode. InGaAs photodiodes are often used.

Edge-emitting lasers operate under high power conditions. Radiation damage increases the power level required to reach the threshold for laser operation. Typical applications use a feedback circuit to increase drive current in order to maintain steady light output. There are two concerns. First, radiation degradation may produce a condition that will affect reliability unless the increase in current is carefully limited to avoid operation at a power level that is too high. Unlike LEDs, where radiation degradation simply decreases optical power, radiation degradation forces higher operating currents in lasers. Second, radiation damage in the control circuit, including the monitor diode, may affect laser performance.

For applications requiring high reliability it is important to operate lasers well below their maximum power level, because the power level will have to increase to accommodate radiation damage and "wearout" effects from normal operation. The usual practice is to limit the allowable change from radiation damage to the range of 10–20%, along with careful thermal control to ensure that the operating temperature of the laser is within narrow bounds. Fortunately, laser diodes are relatively resistant to radiation damage. However, it is important to realize that much closer controls on their operating margins are required for reliable use in space.

11.4 VCSELs

Vertical cavity semiconductor lasers (VCSELs) have different output characteristics compared to conventional stripe lasers. The compact structure and multimode operation of a typical VCSEL affects the optical power characteristics. Internal heating causes the slope efficiency of a VCSEL to become nonlinear at currents that are about three times greater than the threshold current.

There are also pronounced "kinks" in the output characteristic curves, caused by slightly different threshold conditions for the many layers that comprise the distributed feedback grating. The kinks make it more difficult to define threshold conditions, and can vary considerably between different samples from a given lot of devices. They also make it difficult to use derivative plots to evaluate these devices. Figure 11-19 shows how the optical power characteristics of a 950 nm VCSEL are affected by high-energy protons [15].

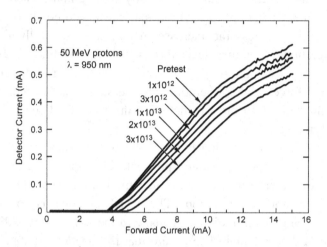

Fig. 11-19. Output power characteristics of a 950 nm VCSEL after exposure to various levels of 50-MeV protons.

Changes in slope efficiency are far more important than for conventional edge-emitting lasers [16]. VCSEL operation must be maintained within a more narrow operating range in order to avoid degradation that could affect reliability. However, radiation test results on several types of VCSELs shows that they are relatively resistant to radiation damage, as is the case for conventional edge-emitting lasers. Note in this example that the kinks in the output characteristics change slightly after irradiation.

Vertical cavity semiconductor lasers are also sensitive to injection-enhanced annealing. Annealing progresses far more rapidly when devices operate above threshold compared to stripe lasers. This may be related to the higher internal temperatures that also cause slope efficiency to be lower for VCSELs. However, damage in VCSELs is a more complex process, affected by the Bragg reflector as well as the resistivity in the cylindrical region of the cavity [17].

Fundamentally, VCSELs have about the same overall sensitivity to radiation damage as edge-emitting lasers. However, the limited region of operation that is caused by the "rollover" at moderate currents must be considered in developing application limits. Unlike edge-emitting lasers, VCSELs are usually used in a narrow current range which reduces some reliability concerns.

11.5 Photodetectors

11.5.1 Conventional photodetectors

The direct bandgap of compound semiconductor detectors makes them relatively insensitive to diffused charge collection compared to silicon detectors. Carrier removal, which can affect heterojunction barriers, is usually the dominant degradation mechanism. However, lifetime damage may affect charge collection near the periphery. It is difficult to predict the damage threshold for compound semiconductor detectors without detailed knowledge of the structure. A comparison of proton damage in an InGaAs detector with a conventional silicon p-n detector is shown in Fig. 11-21. The InGaAs detector degrades at a radiation level that is about two orders of magnitude higher than the silicon detector. Similar results were reported by Marshall and Dale [18].

As discussed in Chapter 4, the wavelength and absorption depth of InGaAs depend on composition. Unpublished results show that the spectral width of the 1300 nm detectors is affected by radiation damage, but that the responsivity at the peak wavelength is less affected than regions away from the peak. Results for detectors optimized at 1300 nm are not necessarily applicable to other InGaAs detectors, and more research is needed to fully understand the mechanisms involved, as well as the dependence of damage on composition.

Although optical sensitivity is usually the most important detector parameter, increases in dark current and noise can also be significant. Those parameters usually degrade less for compound semiconductor detectors compared to silicon detectors because the direct bandgap reduces the total volume of the detector. However, surface leakage can be important, and depends on the specific fabrication method and design.

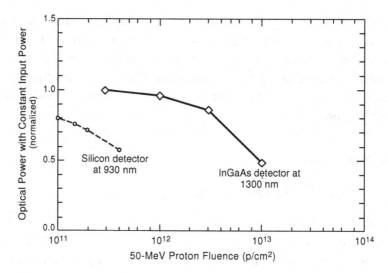

Fig. 11-20. Comparison of proton damage in an InGaAs detector, measured at 1300 nm, with damage in a conventional silicon detector, measured at 930 nm.

11.5.2 Avalanche photodiodes

InGaAs avalanche photodiodes have much lower leakage current compared to Ge photodiodes for low-noise or photon counting applications at longer wavelength. Compound semiconductor APDs have very involved structures, and are somewhat difficult to analyze. Construction information is often proprietary, and very difficult to determine from cross-sectional or other diagnostic methods because many of the layers are extremely thin (see Fig. 4-26).

Recent radiation results for two InGaAs APDs are compared with a germanium APD in Fig. 11-20 [18]. Initial values of dark current and multiplication factor are shown for each device. The Ge device had much higher initial dark current. Despite the much lower initial dark current, one of the InGaAs devices exhibited even larger increases in dark current than the Ge device. The other was far more resistant to damage.

The reason for this large difference in sensitivity has not been fully determined because of limited information about device design and construction. Significant increases in leakage current can occur at proton fluences $<10^{10}$ p/cm^2, making these devices extremely sensitive to radiation damage. Comparisons of cobalt-60 and proton damage show that displacement damage is the dominant damage mechanism for the APDs. However, these devices appear to be an exception to the general observation that compound semiconductors are relatively immune to radiation damage.

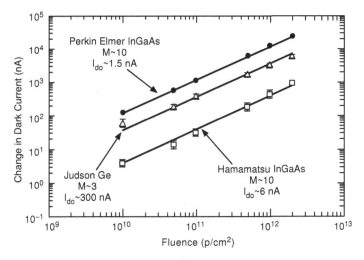

Fig. 11-21. Increase in dark current for three different APD devices after irradiation with 50-MeV protons [19].

11.6 Summary

This chapter has discussed permanent damage in optoelectronic devices. Optoelectronic devices have a much wider range of radiation sensitivity than other classes of compound semiconductors. Some types of light-emitting diodes are extremely sensitive to displacement damage effects, and have been identified as the cause of several failures in space systems. This extreme sensitivity is due to their fabrication method, which requires relatively long lifetimes for efficient operation.

More advanced LEDs, as well as laser diodes, use compact double-heterojunction construction that is far less sensitive to changes in minority carrier lifetime. They operate at higher carrier densities than amphoterically doped LEDs. Both factors reduce their sensitivity to displacements damage effects by an order of magnitude or more.

Annealing effects are very important in optoelectronic devices. Annealing can be an advantage in applications, because a significant amount of the displacement damage may recover for devices that are continuously operated during or after radiation exposure. However, annealing of radiation damage can introduce a great deal of confusion in interpreting radiation test results, and may cause the net effect of radiation damage to be underestimated unless the test conditions closely approximate the actual use conditions.

In contrast to LEDs, laser diodes are far more resistant to radiation damage and also exhibit much less variability between different material systems. This is partly because the bimolecular recombination coefficient has roughly the same value for AlGaAs, InGaAs, and InGaAsP. Laser diodes are strongly affected by annealing, which can be an interference when measuring their radiation performance. However, it is potentially an advantage in applications.

It is important to realize that lasers require very high carrier densities. Even higher carrier densities are required in damaged devices, which can compromise reliability. For that reason it is common practice to limit the maximum increase in threshold current to relatively small values, usually less than a factor of two. Temperature is also an important factor when laser diodes are used in high-reliability applications.

Finally, the direct bandgap of conventional InGaAs photodetectors causes them to be far less susceptible to displacement damage compared to silicon detectors (although the wavelength range is different for the two technologies). However, that improvement does not necessarily extend to avalanche photodiodes, which in some cases are actually worse from the standpoint of radiation damage than germanium APDs used at similar wavelengths.

References

1. B. H. Rose and C. E. Barnes, "Proton Damage Effects on Light Emitting Diodes", J. Appl. Phys, **53**(3), pp. 1772–1780 (1982).
2. B. D. Evans, H. E. Hager and B. W. Hughlock, "5.5 MeV Proton Irradiation of a Strained Quantum-Well Laser Diode and a Multiple Quantum Well LED", IEEE Trans. Nucl. Sci., **40**(6), pp. 1645–1654 (1993).

3. A. H. Johnston, T. F. Miyahira and B. G. Rax, "Proton Damage in Advanced Laser Diodes", IEEE Trans. Nucl. Sci., **48**(6), pp. 1764–1772 (2001).
4. A. H. Johnston, "Radiation Effects in Light-Emitting and Laser Diodes", IEEE Trans. Nucl. Sci., **50**(3), pp. 689–703 (2003).
5. O. Gillard, "Theoretical Study of Radiation Effects on GaAs/AlGaAs and InGaAsP/InP Quantum-Well Lasers", J. Appl. Phys., **93**(4), pp. 1884–1888 (2003).
6. A. L. Barry, et al., "The Energy Dependence of Lifetime Damage Constants in GaAs LEDs for 1–500 MeV Protons", IEEE Trans. Nucl. Sci., **42**(6), pp. 2104–2107 (1995).
7. A. H. Johnston and T. F. Miyahira, "Hardness Assurance Methods for Radiation Degradation of Optocouplers", IEEE Trans. Nucl. Sci., **52**(6), pp. 2649–2656 (2005).
8. A. H. Johnston, et al., "Proton Damage of Light-Emitting Diodes", IEEE Trans. Nucl. Sci., **46**(6), pp. 1781–1789 (1999).
9. A. H. Johnston and T. F. Miyahira, "Energy Dependence of Proton Damage in Optical Emitters", IEEE Trans. Nucl. Sci., **49**(3), pp. 1426–1431 (2002).
10. G. M. Swift, et al., "In-Flight Annealing of Displacement Damage in LEDs: A Galileo Story", IEEE Trans. Nucl. Sci., **50**(3), pp. 1991–1997 (2003).
11. H. N. Becker and A. H. Johnston, "Proton Damage in LEDs with Wavelengths above the Silicon Wavelength Cutoff", IEEE Trans. Nucl. Sci., **51**(6), pp. 3558–3563 (2004).
12. S. M. Khanna, et al., "Proton Energy Dependence of the Light Output in Gallium Nitride Light-Emitting Diodes", IEEE Trans. Nucl. Sci., **51**(5), pp. 2729–2735 (2004).
13. A. H. Johnston and T. F. Miyahira, "Radiation Degradation Mechanisms in Laser Diodes", IEEE Trans. Nucl. Sci., **51**(6), pp. 3564–3571 (2004).
14. C. E. Barnes, "Increased Radiation Hardness of GaAs Laser Diodes at High Current Densities", J. Appl. Phys., **45**(8), pp. 3485–34879 (1974).
15. A. H. Johnston, T. F. Miyahira and B. G. Rax, "Proton Damage in Advanced Laser Diodes", IEEE Trans. Nucl. Sci., **48**(6), pp. 1764–1772 (2001).
16. H. Schöne, et al., "AlGaAs Vertical Cavity Surface Emitting Laser Responses to 4.5-MeV Proton Irradiation", IEEE Phot. Tech. Lett., **9**(12), pp. 1552–1554 (1997).
17. A. Kalavagunta, et al., "Effects of 2-MeV Proton Irradiation on Vertical Cavity Surface Emitting Lasers", IEEE Trans. Nucl. Sci., **50**(6), pp. 1982–1990 (2003).

18. P. W. Marshall and C. J. Dale, "Space Radiation Effects on Optoelectronic Materials for a 1300 nm Fiber Optic Data Bus", IEEE Trans. Nucl. Sci., **39**(6), pp. 1982–1989 (1992).
19. H. N. Becker and A. H. Johnston, "Dark Current Degradation of Near Infrared Avalanche Photodiodes from Proton Irradiation", IEEE Trans. Nucl. Sci., **51**(6), pp. 3572–3578 (2004).

Chapter 12

Radiation Damage in Optocouplers

12.1 Introduction

Despite their simplicity, optocouplers have turned out to be one of the most challenging types of components to deal with in space environments [1–9]. This is due to the "end to end" way in which the electrical properties are specified, with only limited information provided about the optical emitter and light path. Even the emitter wavelength is often omitted.

A diagram of a basic optocoupler is shown in Fig. 12-1 where a simple phototransistor is used to collect the light from an internal light-emitting diode. The most important parameter for an optocoupler is the current transfer ratio (CTR), defined as the ratio between the output current of the phototransistor and the input current of the LED.

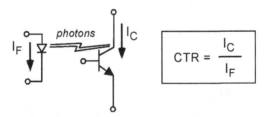

Fig. 12-1. Circuit diagram of a basic optocoupler with a simple phototransistor output.

From a circuit standpoint, CTR is analogous to the common-emitter current gain of a transistor, with a much lower value, typically between 1 and 10. A number of factors influence CTR, including the gain and optical absorption of the phototransistor, properties of the light path, and the optical power emitted by the LED. Their effects are lumped together when the electrical specifications are established, and there is usually no way to evaluate them separately in packaged devices.

The fundamental difference between applications in space and conventional applications of optocouplers is that space radiation can cause such severe degradation in the optical power from the LED that the operation of the optocoupler is shifted into a region where it gradually becomes "starved" for photocurrent. Although wearout phenomena can also affect LEDs (see Chapter 7), far more degradation occurs from space radiation, severely reducing internal operating margins and increasing the effects of unit-to-unit variability on performance.

Some of the issues that affect optocouplers for space applications are listed below:

(1) They are hybrid devices. The LEDs are often purchased from other suppliers without considering the effects of radiation damage, with the possibility that considerably more variability may occur between different production lots compared to conventional components;
(2) LED output decreases with temperature, which must be taken into account during radiation testing and analysis;
(3) Phototransistor characteristics are affected by the optical power level, exacerbating the effect of a decrease in the LED power output;
(4) Physical characteristics of the light path can vary, adding additional uncertainty. In some cases radiation damage may affect light transmission of the light path;
(5) Relatively low electrical signals are involved, which causes some types of optocouplers to be highly sensitive to single-event upset (that topic is addressed in Chapter 13).

The type of LED and wavelength are often unspecified. The wide range of radiation susceptibility of different LED technologies [10-12] makes it nearly impossible to estimate the vulnerability of an optocoupler when the LED type is unknown.

Many different types of optocouplers have been developed, including simple enhancements to the basic phototransistor optocoupler by adding Darlington phototransistors or digital output stages; modifying the basic design by adding a high-sensitivity amplifier; linear optocouplers, which are designed to have a nearly linear relationship between optical input current and output current over an extended current range; and special power switching optocouplers that use a power MOSFET at the output. Those basic types are summarized in Table 12-1, along with a special high-voltage optocoupler with unusual characteristics. High-speed optocouplers (as noted in the shaded row of the table) are unusually sensitive to single-event upset effects. That topic will be addressed in the next chapter.

The material in this chapter will discuss permanent damage in optocouplers, which affects all optocoupler types. We will start with simple phototransistor optocouplers (along with some slight variations), which not only illustrate the issues involved, but have actually failed in space applications. Other optocoupler types will be discussed in later sections, concluding with a power switching optocoupler that fails because of a combination of total dose and displacement effects on its internal components.

Table 12-1. Some Basic Types of Optocouplers.

TYPE	RESPONSE TIME	TYPICAL LED TECHNOLOGY
SIMPLE PHOTOTRANSISTOR	1–2 μs	AMPHOTERIC
DARLINGTON PHOTOTRANSISTOR	VARIES	VARIES
INTERNAL AMPLIFIER	<200 NS	HETEROJUNCTION
LINEAR OPTOCOUPLER	UNSPECIFIED	AMPHOTERIC
POWER SWITCHING OPTOCOUPLER	10 MS	AMPHOTERIC
HIGH VOLTAGE (UP TO 10 KV)	~MS	VARIES

12.2 Damage in Basic Phototransistor Optocouplers

12.2.1 Basic response

The discussion of LED degradation in Chapter 11 pointed out the extreme sensitivity of amphoterically doped LEDs to displacement damage, and it is reasonable to conclude that optocouplers using that type of LED will be dominated by LED degradation. The results in Fig. 12-2 are consistent with this expectation at lower forward current, but not for high forward current. The dashed line shows that the slope is flatter at high currents, particularly at low fluences where the optical power is degrading by less than a factor of two. Although this seems unimportant, it illustrates how optocoupler degradation is affected by the characteristics of the internal phototransistor.

For a simple phototransistor with direct access to the phototransistor base, it is possible to use the phototransistor as a photodiode rather than a phototransistor, eliminating the more complex factors that arise when it is used as a phototransistor.

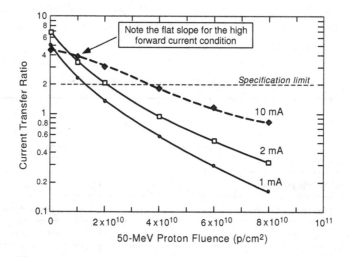

Fig. 12-2. Proton degradation of an optocoupler with an amphoterically doped LED.

Figure 12-3 shows the degradation of CTR when it is interpreted in this way, using $n = 2/3$ (see Eq. 11-2).

The slope is slightly greater than one at higher fluences, but it is almost exactly one when corrections are made to account for the change in diffusion length from radiation damage (the LED wavelength is approximately 890 nm, with an absorption depth of about 30 μm).

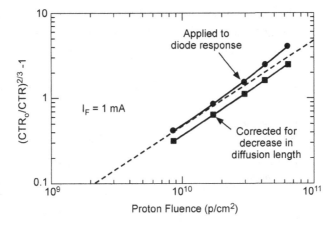

Fig. 12-3. Degradation of an optocoupler when CTR is measured in the photodiode mode.

It is not always possible to get access to the base because the small increase in capacitance from that connection causes a slight loss in frequency response. For that reason, some optocouplers omit the base lead. However, other advantages are provided when the base lead can be accessed separately. Pre-irradiation measurements of the photodiode response provide a baseline value for the amount of optical power that is actually present at the phototransistor, and can be used to ensure that the CTR of a weak LED is not simply being boosted by high phototransistor gain into an operating region where it will pass the end-to-end CTR requirements. Phototransistor properties can also be measured separately. These measurements can be used to eliminate optocouplers with abnormal characteristics, and reduce the range of variability after parts are damaged from radiation.

For optocouplers that use LEDs that are more resistant to radiation damage, the response to radiation is more complicated because LED degradation is no longer the dominant mechanism. Figure 12-4 shows the response of a "radiation tolerant" optocoupler that uses a less sensitive LED, in this case an LED with a wavelength of 660 nm. The shorter wavelength essentially eliminates the need to correct for diffusion length. Photodiode measurements provide an almost ideal straight line (with $n = 0.9$) as shown by the dashed line, just as for discrete LEDs of this type. On the other hand, when we use CTR as a measurement parameter the response is highly nonlinear. The nonlinearity is caused by various effects in the phototransistor, discussed in the next subsection.

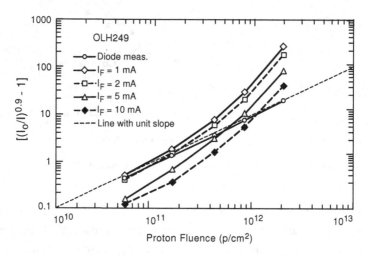

Fig. 12-4. Degradation of a radiation-tolerant optocoupler comparing the diode and phototransistor measurement parameters.

12.2.2 Phototransistor effects

Two different effects arise from the phototransistor: radiation degradation of the phototransistor and its associated diffusion length; and the dependence of phototransistor gain on injection level. The interaction of these factors is not that straightforward.

Figure 12-5 shows the results of a study where an external photodetector was use to measure the degradation of the LED separately from the phototransistor by drilling a small hole in the side of the package [3]. The irradiations were extended to high fluences where the CTR had degraded to less than 1% of initial value.

The nominal wavelength of the LED was 870 nm. A complementary experiment was done to measure photoresponse with an external 870 nm LED. A constant power level was used for the LED.

It is clear from this figure that the optocoupler degradation is much larger than the value that would be calculated from the product of

LED and photoresponse alone. Although there is a slight change in phototransistor gain (not shown in the figure), the main reason for the enhanced degradation of CTR is the lower effective current gain of the phototransistor when it is driven at lower optical power levels by the severely degraded LED.

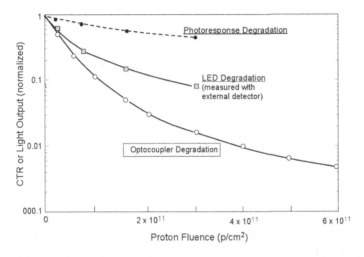

Fig. 12-5. Independent evaluations of photoresponse, LED degradation and CTR showing that CTR degrades far more than expected from the degradation of the other two factors. The minimum specified value of CTR is 2 for a forward LED current of 1 mA.

Note that as the LED degrades – and in this case it is degrading more than a factor of 10 – the phototransistor has to operate at very low power levels that were not considered in the initial optocoupler design by the manufacturer. If we examine the effect of phototransistor gain on base current (an electrical measurement, not a photo-excited measurement) we see how large this effect can be.

Figure 12-6 shows the dependence of current gain on collector current for two phototransistors from one lot of 4N49 optocouplers, prior to irradiation [13]. The gain of the phototransistor in the lower curve will decrease about 30% if we reduce the operating current an order of magnitude. This current related effect will add to the effects of LED degradation, causing CTR to degrade more severely than expected from

LED degradation alone, even if the phototransistor is unaffected by radiation. Any degradation of the phototransistor gain from radiation damage will decrease the CTR even further.

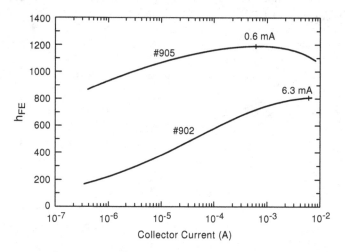

Fig. 12-6. Dependence of common-emitter current gain on collector current for two phototransistors from the same lot of a 4N49 optocoupler. For one device, the normal operating region (without considering radiation degradation) is above the current where the gain is a maximum. The nominal operating region is above 2 mA prior to irradiation.

The upper curve is more interesting. In this case the phototransistor operates above the current where the gain is a maximum (for an undamaged device). This has several consequences. First, it means that the phototransistor gain will actually increase slightly when the LED first degrades because the phototransistor gain is higher with reduced base drive.

This can cause abnormally low changes in CTR at low fluences, which increases unit-to-unit variability, because other phototransistors in the lot behave like the device in the lower curve. Second, as discussed in the next subsection, it tends to flatten the temperature coefficient of CTR in the device specifications. Third, the transistor has a lower dependence on operating current which will cause a device using such a phototransistor to degrade less than other devices in the same lot, even if the LED degradation is the same.

For optocouplers with LEDs that are less affected by radiation damage, the effects of the phototransistor are even more important. Figure 12-7 compares the degradation of a 4N49 optocoupler (with an amphoterically doped 890 nm LED) with that of a similar optocoupler that uses an AlGaAs LED with shorter wavelength.

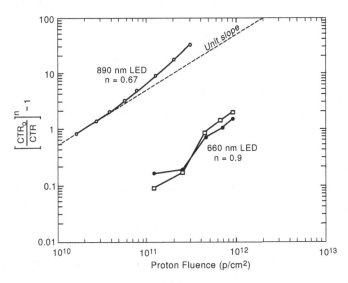

Fig. 12-7. Damage in two different types of optocouplers showing strong nonlinear behavior for a device with shorter wavelengths. The values of n used in the analysis will result in a nearly straight lime when used for LEDs, but does not work very effectively for the optocouplers because of effects in the phototransistors.

The device with the 660 nm LED operates well beyond the peak in the gain-current relationship. Consequently, very little degradation occurs in CTR for the first two data points, but there is a sudden increase in the amount of damage that occurs at higher fluences. The difficulty here is that of interpreting the results, which are highly nonlinear. Although the nonlinearity is clearly evident in this figure, the radiation levels used for the tests are closely spaced, and extend to the point where sufficient radiation damage has taken place to determine the trends in radiation damage at high fluence. In many cases radiation testing is done with fewer radiation levels, where the nonlinear nature of the damage is less evident. The optocoupler with the improved LED clearly performs better in a radiation environment, but the strong nonlinearity in the degradation characteristics make it far more difficult to assign post-radiation values to design parameters.

Another effect becomes important for phototransistors when an LED with long wavelength is used. When the effective base current is produced from the collection of photo-induced carriers, more degradation takes place compared to the degradation that is measured using an external base current. This effect is particularly important at long wavelengths, because part of the photocurrent is collected by diffusion, which is reduced by radiation damage because the minority carrier lifetime degrades. Figure 12-8 illustrates this difference, which becomes important at high fluences. The experiment was done by using external light sources – using a constant light intensity – at the wavelengths shown in the figure.

The additional degradation from this effect will increase the importance of phototransistor degradation. It is particularly important when radiation-tolerant LEDs with longer wavelengths are used. In practice, more degradation will take place than indicated by experiments from an external source because light from the internal LED does not strike the surface of the phototransistor at normal incidence. Thus, the experimental result is only an approximation.

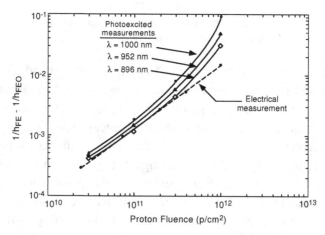

Fig. 12-8. Evaluation of phototransistor damage comparing electrical measurements with measurements using photo-excitation from external light sources with various wavelengths.

12.2.3 Effect of different particle types

Even though it was discussed in Chapter 9, it is worth revisiting the effects of different particles on degradation of optocouplers for two reasons: first, displacement damage measurements of optocouplers with amphoterically doped LEDs do not agree with the calculated values of non-ionizing energy loss for protons with high energies or for neutrons [14,15]; and second, we have to consider damage in both the LED and detector, which are usually fabricated with different materials. If both components contribute to the damage, we need to avoid making damage comparisons in regions where damage equivalence through NIEL calculations disagrees with experiment.

Figure 12-9 compares the degradation of optocouplers from two different manufacturers when they are tested with cobalt-60 gamma rays, 1-MeV (equivalent) neutrons from a nuclear reactor, and 200 MeV protons. First, note that almost no damage takes place when gamma rays are used, emphasizing once again that gamma ray testing is incorrect for this type of component. Second, note the difference between neutron and proton damage, even though calculations of non-ionizing energy loss predict almost the same amount of damage for those two radiation sources.

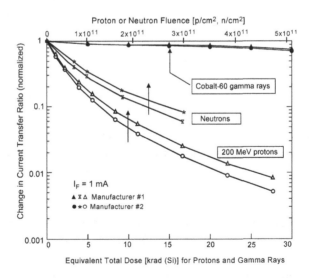

Fig. 12-9. Effect of different particle types on damage in a widely used optocoupler with an amphoterically doped LED.

The disparity between NIEL calculations and experimental results point out the need to use particle types and energies that are representative of the actual environment. For space applications, testing should be done with protons or electrons, depending on which type of particle produces the most displacement damage. For earth-orbiting spacecraft in low-earth orbits, protons usually dominate, whereas electron damage dominates for geosynchronous orbits.

12.2.4 Temperature effects

Even modest temperature changes can affect optocoupler performance. LED temperature dependence is straightforward, ranging from -0.7 to -1.1% per degree Celsius. However, for a transistor that operates above the current where gain is maximum, that temperature sensitivity can be masked by the characteristics of the phototransistor (see Fig. 12-6).

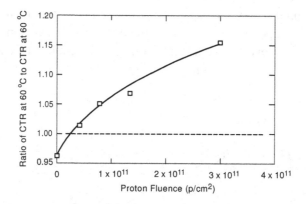

Fig. 12-10. Change in temperature sensitivity of an optocoupler after proton irradiation. The increase in temperature sensitivity of CTR is caused by changes in the operating characteristics of the phototransistor that mask temperature effects for unirradiated devices when the phototransistor operates above the current where gain is a maximum.

For example, Fig. 12-10 shows how the temperature dependence of an optocoupler changes with increasing levels of degradation. Initially CTR has a slightly positive temperature dependence, even though the temperature dependence of the internal LED is negative. As the operating point shifts (due to LED degradation), the temperature dependence of the CTR increases. Although the 16% decrease in CTR at 60°C appears small, the temperature of conduction cooled circuit boards can be well above the nominal spacecraft back plane temperature The effect of the temperature dependence adds to other factors (including the distribution of CTR values within a production lot), further decreasing the optocoupler operating margin in the application.

12.2.5 Production lot variability

Due to the hybrid construction, optocouplers are potentially more prone to production lot and unit-to-unit variability compared to other components. Variation in the radiation response of the LED is usually the most important contributor to such variations. Figure 12-11 compares degradation of current transfer ratio of the 4N49 optocoupler for three different production lots. All are from a single manufacturer, and meet the same high-reliability requirements. The figure shows mean values for each lot. The fluence where the devices no longer meet their minimum requirements differs by about a factor of four. Although 50-MeV protons were used to evaluate the devices, the equivalent total dose level of the most sensitive lot is below 2 krad, a very low radiation level.

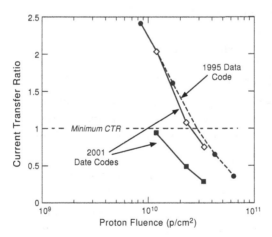

Fig. 12-11. Comparison of degradation of the current transfer ratio of devices from one manufacturer. The variability is caused by differences in the LEDs used by the manufacturer.

Variations in the light path can also affect optocoupler uniformity. One type of optocoupler uses a silicone compound that surrounds the LED and phototransistor, essentially serving as a crude light pipe (most of the light strikes the interface between the coupling compound and nitrogen within the package at an angle that exceeds the Brewster angle, resulting in total internal reflection). Voids (air bubbles) within the silicone compound are allowed by the specifications, which reduce light coupling efficiency, and add to unit-to-unit variability in performance. Other physical factors that can affect performance include roughness at the edge of the LED (where most of the light is emitted), and inconsistencies in the physical placement of the LED and phototransistor.

There are various ways to deal with the inherent variability of optocoupler degradation. The first is to use very conservative designs that can still function after the optocoupler has degraded severely. However, this may be difficult for optocouplers with amphoterically doped LEDs because they degrade so much at very low radiation levels. Another approach is to use a combination of lot sample radiation testing along with electrical screening of all devices within the lot to eliminate any devices with CTR values that are close to the minimum limit.

12.3 Optocouplers with High-Speed Internal Amplifiers

12.3.1 Operational characteristics

A high-speed amplifier can be used instead of a phototransistor to improve the response time of an optocoupler. The LED must also have a fast response time. The design requires a tradeoff between the requirement for high speed, and the reduced light output of high-speed LEDs (see Chapter 4).

Figure 12-12 shows a diagram; the photodiode is integrated with the amplifier in order to reduce parasitic capacitance and improve the frequency response. In order to ensure compatibility with the semiconductor process used for the amplifier, the detector typically has higher doping levels and much shorter light collection depth compared to optocouplers with phototransistors.

Radiation Damage in Optocouplers 325

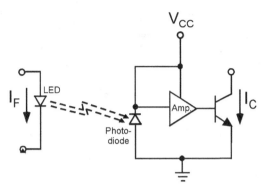

Fig. 12-12. Diagram of a high-speed optocoupler with an internal amplfier.

A major manufacturer of these device types uses GaAsP LEDs with a nominal wavelength of 700 nm. Some variants have a digital output stage instead of the open-collector configuration shown above, providing a digital output voltage instead of current. The conventional definition of CTR cannot be used with optocouplers with digital output stages. It is also limited for the circuit in Fig. 12-12 by the performance characteristics of the high-gain amplifier, which will stop functioning if the optical input power from the LED is too low.

For the digital output case, it is possible to measure the transfer characteristics by sweeping the LED input current until the output switches from high to low. In cases where hysteresis is added, there are two possible transition conditions, with slightly different input current requirements due to the hysteresis.

For either the analog or digital output case the chief concern for permanent damage is the point where catastrophic failure occurs. Fortunately the GaAsP LEDs used for these devices require much higher radiation levels before significant degradation takes place, decreasing the importance of the catastrophic failure mode until very high radiation levels are reached.

Figure 12-13 shows proton degradation of a 6N134 optocoupler using the analysis method developed in Chapter 11. The degradation is nearly linear up to about 10^{13} p/cm^2. An analog measurement of the threshold condition was made by sweeping the input current until the output current abruptly increased (not a standard measurement).

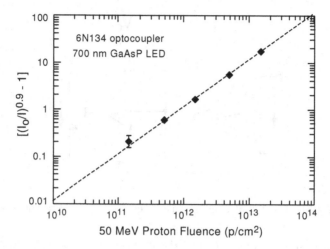

Fig. 12-13. Degradation of an optocoupler with a high-speed amplifier.

The data in Fig. 12-13 are representative of typical devices from this particular lot. However, one of the 16 parts within this lot exhibited catastrophic failure when the device was irradiated above 5×10^{12} to the higher fluence of 10^{14} p/cm^2.

This is shown in Fig. 12-14, along with two other devices that are representative of the other devices from the same lot. The device that failed did not operate even when the input current was increased to 30 mA, a level that is about 15 times above the nominal switching point of a fresh device. This implies that the internal amplifier was degraded to the point where it was no longer functional.

We need to contrast this behavior with that of phototransistor-based optocouplers. Severe degradation can take place in optocouplers with simple phototransistors, but the dominant effect is a sharp reduction in current transfer ratio. Those devices will still function, even when the CTR is degraded by more than two orders of magnitude (see Fig. 12-8). Although amplifiers with internal amplifiers are far more resistant to radiation damage, the catastrophic failure mode could be extremely important. Tests of these types of optocouplers must be extended to radiation levels that are well above radiation levels anticipated in the actual environment, carefully examining device performance and variability to ensure that catastrophic failure will not take place.

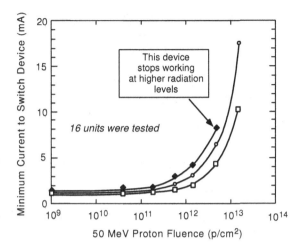

Fig. 12-14. Catastrophic failure of one of sixteen units of an optocoupler with an internal high-gain amplifier. The failure is most likely caused by damage to the internal amplifier.

12.4 Optocouplers with MOSFET Output Stages

Although it is not possible to discuss all types of optocouplers, optocouplers with power MOSFET output provide an interesting example of an even more complex optocoupler function [16]. Figure 12-15 shows an example of the implementation, which used three different elements: an LED, mounted on a separate substrate that is placed vertically over the top of the other components; a special photovoltaic chip that develops a high forward voltage when the LED is turned on; and one or more discrete power MOSFETs at the output, which are switched on from the photovoltaic chip.

The stack of photodiodes functions in much the same way as a solar array. One indirect effect of this photodiode stack is extremely slow switching speed. Most of the photocurrent is collected by diffusion, a relatively slow process. However, the turn-off speed is even slower because of the high effective resistance of the photodiode string after the LED is turned off. As a result these optocouplers require time periods >10 ms for switching, which is slow enough to affect electrical measurements unless the integration time used for digital readings is extended to comparable time periods.

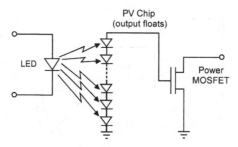

Fig. 12-15. Circuit diagram of the components used within a typical power switching optocoupler with MOSFET output.

Unlike other types of optocouplers, those with power MOSFET are degraded by total dose (which affects the MOSFET) as well as displacement damage (which primarily affects the LED). This is an added complication both for radiation testing and for applications. Bias conditions during irradiation have a large effect on the total dose mechanism, but little effect on the LED.

The total dose mechanism can be evaluated separately by irradiating samples with gamma rays, essentially eliminating damage from the LEDs. Figure 12-16 shows an example, where the gate of the MOSFET was biased by turning on the LED during the irradiation.

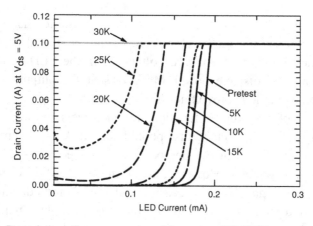

Fig. 12-16. Degradation of an optocoupler with a power MOSFET output, using gamma rays. Because total dose damage is sensitive to bias conditions, the LED was forward biased during the irradiation to provide a positive gate voltage on the MOSFET.

The minimum current required to switch the power MOSFET decreases with increasing total dose because the threshold voltage becomes more negative. However, after approximately 20 krad(Si) the subthreshold leakage current of the power MOSFET becomes high enough so that the device is no longer fully off. At a total dose of 30 krad(Si) the leakage current has increased above 100 mA, which was the compliance limit of the instrumentation used in this particular test.

When a similar test is done using protons, the results are due to a combination of the increasing leakage current of the power MOSFET and the decrease in the light output of the LED (recall that protons produce ionization damage as well as displacement damage). Figure 12-17 shows a typical result. At low proton fluence there is little degradation in the LED, and power MOSFET total dose effects dominate. As the fluence increases, higher LED currents are needed to switch the device, but the leakage current of the power MOSFET also continues to increase. The competing effects of these mechanisms cause the threshold LED current to decrease at low levels, but increase at higher levels, as shown in the figure. Subthreshold leakage current in the MOSFET is the dominant failure mode at higher radiation levels, causing the device to be "on" regardless of the LED input current.

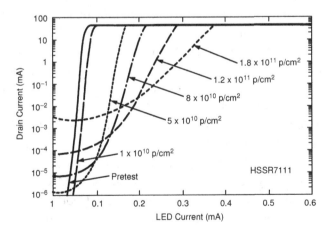

Fig. 12-17. Degradation of the same type of optocoupler used in Fig. 12-16 when the device is irradiated with protons instead of gamma rays. The minimum LED current initially decreases, but then increases due to the competing effects of total dose damage in the MOSFET and displacement damage in the LED.

The power MOSFET optocoupler is a good example of the complex radiation response of specialized optocouplers. There are several competing effects, which depend on the particular type of component that is used internally, as well as the way that the device is biased during the time that it is irradiated. The catastrophic failure mode of the power MOSFET depends on bias conditions during irradiation, and could be overlooked if the part was tested without the voltage conditions that are necessary for this mode to occur.

The device specifications did not provide sufficient detail about the way that the device is constructed to allow a technically sound plan for testing and evaluation. Only the end-to-end functionality was included. It was necessary to partially disassemble one of the devices in order to determine how it functioned, and how the various internal components interacted. A similar approach may be needed for other types of optocouplers with elaborate designs, and do not neatly fit into the two categories of basic optocouplers that were discussed earlier.

12.5 Summary

The combination of silicon-based components for light detection and amplification and compound semiconductors for light emission makes it particularly challenging to evaluate optocouplers for use in space. Although many optocouplers are designed in a straightforward manner, it is possible to develop devices with far more complexity, such as the power MOSFET optocoupler discussed above. A basic understanding of the optocoupler design and the way that the various internal components interact is essential in order to evaluate them for space applications.

The sensitivity of optocouplers to damage from space radiation various over an exceptionally wide range, just as for light-emitting diodes. This is partly due to the different types of LED technologies that are used, but the properties of the silicon-based components are also important. It is important to understand the need for testing optocouplers with protons or electrons instead of gamma rays.

The particle type and energy must be based on the actual space environment for the application. The dependence of non-ionizing energy loss must be taken into account in order to use test results at a single particle energy to calculated degradation in the actual environment, and must consider effects in other components (usually silicon) as well as the LED.

Finally, some types of optocouplers have catastrophic failure modes that will likely have far more impact on their performance in space compared to the gradual degradation of basic optocouplers with phototransistor output stages. For the power MOSFET optocoupler, the radiation level and specific failure modes depend on the bias conditions during the time that it is irradiated. More thorough evaluations of optocouplers with catastrophic failure modes are needed in order to ensure that they will operate satisfactorily.

References

1. K. Soda, C. Barnes and R. Kiehl, "The Effects of Gamma Radiation on Optical Isolators", IEEE Trans. Nucl. Sci., **22**(6), pp. 2275–2279 (1975).
2. H. Lischa, H. Henschel, O. Kohn, W. Lennartz and H. Schmidt, "Radiation Effects in Light-Emitting Diodes, Photodiodes, and Optocouplers", Proc. RADECS Conf., 1993, pp. 226–231.
3. B. G. Rax, C. I. Lee, A. H. Johnston and C. E. Barnes, "Total Dose and Proton Damage in Optocouplers", IEEE Trans. Nucl. Sci., **43**(6), pp. 3167–3173 (1996).
4. M. D'Ordine, "Proton Damage in Optocouplers", IEEE Radiation Effects Data Workshop, pp. 122–124 (1997).
5. R. A. Reed, et al., "Emerging Optocoupler Issues with Energetic Particle-Induced Transients and Permanent Radiation Degradation", IEEE Trans. Nucl. Sci., **45**(6), pp. 2833–2840 (1998).
6. A. H. Johnston and B. G. Rax, "Proton Damage in Linear and Digital Optocouplers", IEEE Trans. Nucl. Sci., **47**(3), pp. 675–682 (2000).
7. K. A. LaBel, et al., "A Compendium of Recent Optocoupler Radiation Test Data", IEEE Radiation Effects Data Workshop, pp. 123–146 (2000).
8. R. Mangeret, et al., "Radiation Characterization and Test Methodology Study of Optocouplers for Space Applications", paper C-5, presented at the 2001 RADECS Conference, Grenoble, France, September, 2001.
9. R. Germanicus, et al., "Analysis of the Proton Induced Permanent Damage in an Optocoupler", IEEE Trans. Nucl. Sci., **49**(3), pp. 1421–1425 (2002).

10. A. H. Johnston and T. F. Miyahira, "LED Technologies for Optocouplers", IEEE Trans. Nucl. Sci., **54**(6), pp. 2112–2220 (2007).
11. L. R. Dawson, "High Efficiency Graded Bandgap AlGaAs Light-Emitting Diodes", J. Appl. Phys., **48**(6), pp. 2485–2493 (1977).
12. B. H. Rose and C. E. Barnes, "Proton Damage Effects on Light Emitting Diodes", J. Appl. Phys., **53**(3), pp. 1772–1780 (1982).
13. A. H. Johnston and T. F. Miyahira, "Hardness Assurance Methods for Radiation Degradation of Optocouplers", IEEE Trans. Nucl. Sci., **52**(6), pp. 2649–2656 (2005).
14. A. L. Barry, *et al.*, "The Energy Dependence of Lifetime Damage Constants in GaAs LEDs for 1–500 MeV", IEEE Trans. Nucl. Sci., **42**(6), pp. 2104–2107 (1995).
15. R. A. Reed, *et al.*, "Dependence of Proton Damage Energy in AlGaAs Light-Emitting Diodes", IEEE Trans. Nucl. Sci., **47**(6), pp. 2492–2499 (2000).
16. A. H. Johnston and T. F. Miyahira, "Radiation Damage in Power MOSFET Optocouplers", IEEE Trans. Nucl. Sci., **54**(4), pp. 1104–1109 (2007).

Chapter 13

Effects from Single Particles

This chapter discusses the effects of short-duration currents produced by charged particles on compound semiconductor devices, optocouplers and optical receivers. The currents are caused by the ionization track created by a single charged particle along its path through a semiconductor or insulator. If the charge from the ion strike is large enough, it may disrupt the normal function of a circuit, producing a transient effect that can alter the device.

The location of the ion strike within a device (or circuit) is also important. It has to strike near a sensitive region in order to affect the device. The effective area for such effects is defined by the cross section (usually determined experimentally), and plotted vs. the ionization strength of the particle.

Galactic cosmic rays are usually one of the most important environments when we consider these effects. Even though the cosmic ray fluence is much lower than for protons and electrons, the fact that only one particle is involved causes single-event effects to be important at much lower fluences compared to the integrated damage effects that were discussed in Chapters 10–12.

In most cases these small charges produce extraneous "noise like" currents, resulting in only a temporary circuit malfunction that spontaneously recovers (single-event upset, or SEU). However, for some devices permanent damage takes place – *e.g.*, burnout of the semiconductor, or rupture of a gate or other insulator. Permanent damage effects from single particles are well known for silicon technology, but have only recently been observed in compound semiconductors.

13.1 Basic Concepts

13.1.1 Charge deposition

As discussed in Chapter 8, a charged particle produces an intense track of electron-hole pairs along its path when it passes through a semiconductor or insulator. The diameter of the initial charge track is approximately 0.1 µm. It increases to about 1 µm approximately 1 ns after the particle strike, due to ambipolar diffusion. The charge density is very high, distorting the electric field along and near the track of the particle. This is illustrated in Fig. 13-1, where the particle strikes a reverse-biased n-p junction. The dashed lines show the junction depletion layer, which collapses along the particle strike because of the very high charge density. In this example the doping density is lower in the p-region, causing more field distortion in that region compared to the n-region.

Not all of the charge produced by a particle strike is effective in causing effects in devices. Some fraction of the charge will be collected when it traverses (or strikes near) a p-n junction, but part of the charge will either migrate to other regions, or recombine. This is discussed further in Section 13.1.3.

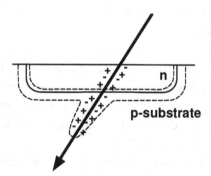

Fig. 13-1. Diagram of charge generation along a particle strike through an n-p junction.

The charge density produced by an ion strike is proportional to linear energy transfer, LET. The spectrum of LET values that we need to be concerned with in the natural space environment was shown previously in Fig. 8-8. The number of ions decreases with increasing LET. It declines gradually up to about 20 MeV-cm^2/mg, but abruptly declines by about four orders of magnitude as we increase LET from 20 to 30 MeV-cm^2/mg. There are very few particles with LET values >30 MeV-cm^2/mg, which allows us to use that value as a benchmark for SEU sensitivity, although angular effects cause the effective LET to be higher in many cases (see Section 13.1.3). In deep space, the total number of particles with LET >30 MeV-cm^2/mg is about one per square cm per year.

Another process can occur because of secondary reactions between the incoming particle and a lattice atom within the semiconductor (or in surrounding regions) [1]. This process involves *nuclear* collisions rather than direct ionization, as depicted in Fig. 13-2. The cross section for nuclear interactions is about five orders of magnitude lower than for direct ionization from charged particles.

The recoil atom has less energy than the incoming particle, but it can interact with electrons in the target material through ionization instead of nuclear reactions, with a much higher cross section and a higher LET compared to the primary particle. For example, the LET of a 100 MeV proton is ~0.006 MeV-cm^2/mg. The maximum LET of a recoil atom in gallium arsenide from a proton with that energy is about 13 MeV-cm^2/mg for elastic interactions, a factor of 2100 higher than for the LET of the incident proton.

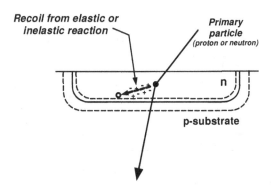

Fig. 13-2. Production of charge by high-energy charged particles the short-range recoil products of nuclear interactions (indirect process).

Although the LET of the recoil atom is higher, it has a short path length, typically only a few microns. Thus, the increased LET of the recoil is offset by the short path length, unless the charge collection depth of the device is small. Upset rate calculations for indirect ionization depend very strongly on the assumptions made about the charge collection path for the device structure, which in many cases can only be estimated. Neutrons with energies >15 MeV produce reactions in semiconductors that are similar to reactions from protons, and produce recoil atoms in an analogous way.

13.1.2 Charge density produced by particles in space

As discussed in Section 8.2, the charge density per unit path length is proportional to LET. LET values for several different particles are shown in Table 13-1 for silicon and GaAs. Note the very low LET for protons. For most devices the proton LET is so low that direct ionization does not contribute to transient upsets. As a result, proton-induced upsets are almost always caused by recoil atoms from nuclear interactions. Even though the nuclear reaction cross section for protons is small, space environments often contain large numbers of high-energy protons. This makes it possible for nuclear recoils from protons to make a comparable (or even higher) contribution to the overall upset rate as the contribution of galactic cosmic rays. The galactic cosmic rays interact by direct ionization.

Table 13-1. Comparison of Charge Densities for Various Particles.

Particle Type	Linear Energy Transfer in Si (MeV-cm^2/mg)	Linear Energy Transfer in GaAs (MeV-cm^2/mg)	Ratio of LET$_{GaAs}$ to LET$_{Si}$
100-MeV iron	29.2	19.43	0.67
100-MeV oxygen	3.11	2.15	0.69
100-MeV alpha	0.068	.053	0.78
6-MeV alpha	0.55	0.36	0.65
100-MeV proton	0.006	.0046	0.77
15-MeV proton	0.025	.019	0.77

The stopping power is different for silicon and GaAs, partly because of the different density of the two materials. The last column in Table 13-1 shows the LET ratio of Ga and silicon. LET in GaAs is 25–35% lower than that of silicon. The ratio changes very little for the various particle types and energies.

13.1.3 Charge collection

As discussed earlier, only a fraction of the charge deposited by an energetic ion will be collected. There are several reasons for this. First, part of the charge within the dense plasma created by the particle will recombine before it can be collected. Second, for very small geometry devices, the size of the node at which charge collection takes place is smaller than the diameter of the charge track. The charge track reaches a quasi-equilibrium condition about 1 ns after it is created, due to ambipolar diffusion. The track diameter at that time is on the order of 1 µm, considerably larger than the collector geometry of a highly scaled, minimum size transistor. Charge generated beyond the node may either recombine without being collected, or may be collected by other active junctions within the circuit. Finally, charge collection will terminate at regions that are very highly doped. This can limit the charge collection length in vertical and lateral dimensions. For example, a SiGe transistor on a bulk substrate can collect charge over a distance >20 µm. In contrast, the charge collection depth on a similar transistor with a highly doped substrate is limited to about 2 µm (effectively the thickness of the lightly doped eptiaxial layer), reducing the collected charge by about one order of magnitude. Device geometry is critically important for charge collection, particularly for highly scaled devices.

The collected charge also depends on the location and angle of the particle strike. More charge will be collected for a normally incident particle near the center of the junction involved in charge collection than for other locations. Models for the effect of geometry on charge collection often assume ideal rectangular charge collection dimensions that do not account for the many different layers and doping levels involved in compound semiconductor devices. For a rectangular structure with a shallow depth and high aspect ratio it is usually assumed that the effective amount of charge that is produced increases with the angle of incidence, *e.g.*,

$$\text{LET}_{\text{eff}} = \text{LET}_{\text{norm}} * \frac{1}{\cos\theta} \quad (13\text{-}1)$$

where LET_{eff} is the effective value of the particle LET when it strikes the charge collection region at a normally incident angle, θ, and LET_{norm} is the value of LET at normal incidence.

This relationship, often referred to as the "cosine law", assumes that the range of the particle exceeds the distance over which the charge is collected. From a practical standpoint it causes the effective LET that we must consider in space environments to be about 75 MeV-cm^2/mg for most SEU phenomena rather than 30 MeV-cm^2/mg, which is the LET where the abundance of galactic cosmic rays falls abruptly.

The applicability of Eq. 13-1 depends on device geometry as well as the charge collection mechanism. For modern parts with small area, it does not apply to devices where most of the charge is collected by diffusion, or where the active area is much smaller than the effective diameter of the particle strike.

13.1.4 Critical charge

The short times involved for charge deposition and collection allow circuit responses to transient events to be evaluated using the concept of critical charge. The critical charge is the minimum *collected* charge at a particular circuit node that will cause the circuit to change state. This is a very useful concept, but it is based on simple circuit concepts that have limited accuracy. It also depends on the fraction of deposited charge that is actually collected by the sensitive node, which can only be estimated. In addition, some structures contain internal transistors that can amplify collected charge, adding an additional layer of complexity to the critical charge concept.

Critical charge is a useful way to examine circuit upset effects, but it has limited value for high-speed devices with small feature size because of the many variables that affect charge collection in real device structures.

13.2 Single-Event Upset in Logic Devices

The fast response times of compound semiconductor logic devices, a major advantage for many applications, also makes them extremely sensitive to upsets from high-energy particles. Figure 13-3 shows a basic flip-flop, using enhancement and depletion mode GaAs JFETs. The arrow shows the location of a particle strike in one of the enhancement-mode JFETs where a short pulse of current can cause the flip-flop to change state. Ion strikes in the "Set" transistor and the lower transistor in the output stage can also produce upsets. The high speed and low noise immunity (~0.15 V) of this circuit makes it inherently sensitive to SEU effects.

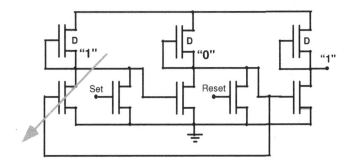

Fig. 13-3. Circuit diagram of a flip-flop using E/D MESFETs.

Earlier GaAs MESFETs were made on semi-insulating substrates, without a buffer layer between the active devices and the substrate. Semi-insulating substrates have extremely long lifetime. Coupling between the substrate and the active devices increased the amount of charge that was collected by the high-speed MESFETs. It also extended the response time to about 1 ms, a surprising result when it was first discovered. Modern MESFET integrated circuits use a buffer layer that eliminates such long-duration responses.

MESFETs and HFETs also contain parasitic *bipolar* transistors. Those parasitic transistors can be turned on by a charged particle strike, amplifying the collected charge to the point where significantly more charge is collected at the circuit node than is deposited by the ion strike. Hughlock, *et al.*, verified this mechanism with special MEFET test structures where the gate was deliberately omitted [2]. The same response occurred for MESFETs with and without the gate structure, verifying that the MESFET was not involved in the charge amplification process, implying that other mechanisms were responsible. The parasitic bipolar structure was the same for both types of test structures.

A comparison of the SEU response of three different logic families, manufactured with a feature size of 1 µm, is shown in Fig. 13-4 [2]. The very low threshold LET of the GaAs MESFET is due to higher bipolar transistor gain for that particular structure, along with additional charge collection from the semi-insulating substrate. The complementary HFET and InP HFET technologies incorporate buffer layers between the active devices and the substrate, eliminating the effect of charge collection in the substrate region. They also have lower bipolar transistor gain. Both factors improve SEU hardness compared to the MESFET.

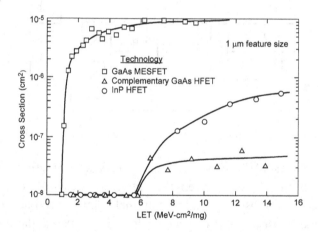

Fig. 13-4. Comparison of single-event upset sensitivity of GaAs MESFETs fabricated on a semi-insulating substrate with other compound semiconductor logic devices [2].

Recent work by McMorrow, *et al.*, showed that excess charge collection also takes place in AlSb/InAs HFETs [3]. Those devices were fabricated with much smaller feature size, 0.1 µm, with a threshold LET of about 1 MeV-cm^2/mg. Such low threshold LETs are typical of advanced compound semiconductor logic devices. They are usually sensitive to upset from protons as well as from heavy ions. Despite the low threshold LET, the cross section is usually low because of the small cell size. The integration density of these circuits is usually very small compared to that of mainstream silicon devices, reducing the overall upset rate when we consider the effect on the complete circuit.

Additional work has been done on SiGe devices, which are more likely to be used in high-density large-scale circuits compared to other compound semiconductor technologies. This increases the practical importance of single-event upsets compared to the small-scale circuits that are typical of other compound semiconductor technologies. For example, a large-scale SiGe circuit could have more than 10^7 transistors with a die area of about 1 cm^2. The upset rate of such a large circuit would be many orders of magnitude higher than that of a small circuit made with other compound semiconductor materials that have defect densities that are too large to make such large circuits.

In many cases single-event upset effects are only a minor annoyance. System methods such as parity checking, or error-detection-and-correction using extended word lengths, can successfully overcome the effects of small numbers of upsets. When using such approaches, it is important to recognize that a single particle strike may cause several different storage cells to upset from a single particle strike (multiple bit upset).

Another way to deal with the high upset sensitivity is through circuit hardening, using a combination of circuit design and device geometry to either eliminate the SEU sensitivity, or at least increase the threshold LET so that the total cross section for upsets is reduced.

The first efforts to develop hardened flip-flops using SiGe processes were disappointing. Long response times were observed in the redundant circuit elements used for hardening that were inconsistent with assumptions used in the design [4]. Subsequent modeling work showed unexpectedly large currents in substrates with high doping, which was the likely cause of the poor performance of hardened SiGe logic devices [5]. Modeling results are shown in Fig. 13-5 for LET = 0.3 pC/μm, a very low value. The peak current for the highly doped structure exceeds 10 mA, even for such a low LET. The current for a structure with low substrate doping density is lower, but extends to longer times. High doping levels shorten response time, but increase prompt current. This makes the SEU problem far worse for such high speed devices.

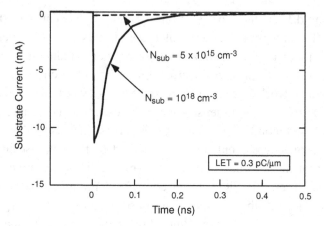

Fig. 13-5. Modeling results showing large substrate current in SiGe transistor structures with high substrate doping levels [4]. © 2001 IEEE. Reprinted with permission.

These examples show that compound semiconductor technologies used for high-speed logic applications are extremely sensitive to single-event upset. This is due to their small geometry and high speed, which reduce the critical charge, along with parasitic bipolar transistors that provide charge amplification.

The available data and modeling results suggests that this trend towards high SEU sensitivity will continue as devices are scaled further. Fortunately most applications of compound semiconductors are in circuits with relatively simple functions and small die size. This reduces the upset probability because there are relatively few circuit elements that can be upset, partially compensating for the low threshold LET.

The small geometry reduces the cross section for upset in a single small area transistor, which makes the upset rate smaller when it is normalized to a single bit. This partially compensates for the low threshold LET provided that there are relatively few components per chip, which is often the case for the specialized components that are typical of compound semiconductors.

SEU effects are less important for devices with larger geometry, such as power amplifiers, because their overall geometry is much larger. The larger geometry increases critical charge, reducing the number of particles that can produce upsets. Even though power MESFETs and

HFETs have individual gate lengths on the order of 0.1 µm, the devices typically consist of many such elements connected in parallel. This increases the critical charge, essentially eliminating the SEU response mechanism compared to small area logic devices with comparable gate lengths.

13.3 Optocouplers

13.3.1 Basic design and sensitivity

Optocouplers are among the most sensitive devices to transient effects from heavy particles or protons. This sensitivity is inherent in the basic design, which was discussed in Section 12.1. Optocouplers that contain a simple phototransistor are less affected by SEU than those containing internal amplifiers with high gain because the response time of a phototransistor is relatively slow, and there is less gain.

The "end to end" efficiency of an optocoupler is relatively low. Typically the conversion efficiency of current through the light emitting diode to photons at the output of the LED is on the order of 1%, and only a fraction of those photons are collected by the photodetector. Consequently optocouplers must be capable of detecting relatively small signals. This makes them inherently sensitive to the spurious charges produced by radiation.

We can calculate the approximate charge sensitivity of an optocoupler from the basic properties of the photodiode and the response time. For example, consider a high-speed optocoupler with a 25 ns response time and input LED threshold current for switching of 3 mA. Assuming a net current conversion efficiency of 0.2% between the input current of the LED and the charge collected by the photodetector in the integrated amplifier, the optocoupler must be capable of amplifying a charge of

$$(3 \text{ mA}) \cdot (25 \text{ ns}) = 75 \text{ fC} \qquad (13\text{-}2)$$

This is a very small charge. Radiation-induced charges of that magnitude would then be expected to produce spurious signals in the optocoupler.

Recall that a particle with an LET of 1 MeV-cm^2/mg produces 10 fC in a path length of only one micrometer. Thus, we expect high-speed optocouplers to have an upset sensitivity near – or below – an LET of 1 MeV-cm^2/mg. This is much lower than the LET threshold of most conventional digital microelectronic devices.

Another factor in the upset sensitivity of optocouplers is the area of the photodetector. It must be large enough to avoid misalignment errors during fabrication. For high-speed optocouplers the LED is usually on a separate (GaAs) substrate that is placed over the silicon phototransistor and amplifier, located on a separate silicon substrate. The photodetector area must be larger than the area of LED light emission because of fabrication constraints. The extra area increases the sensitive area for single-event upset, but does not contribute to light collection.

13.3.2 Transients from high-energy particles

13.3.2.1 Heavy ions

Experimental results for a high-speed optocoupler during a test run with heavy particles are shown in Fig. 13-6. The output of the optocoupler is normally at 5 V, but transients from heavy ions that are produced within either the internal photodiode or the high-gain amplifier cause spurious output signals. The particles strike the device in random locations when it is exposed to particles from the accelerator, which is the reason for the waveform variability. The amplitude and width of the output transients depend on the location of the particle strike. If it occurs near the center of the photodiode the response is higher compared to a particle strike near the periphery of the photodiode, or within the amplifier.

The results of a series of experiments with different particle types and energies are shown in Fig. 13-7 [6]. The cross section is measured with different types of ions, showing how the relative number of events increases for particles with higher specific charge density. From basic geometrical considerations one would expect the cross section to saturate at a value that corresponds to the area of the internal photodiode, but this does not occur in practice because it is possible to collect additional charge by diffusion from particles that strike outside the physical region of the diode. The LET threshold, ~0.3 MeV-cm^2/mg, is extremely low for this type of device, consistent with the estimates provided in the previous subsection.

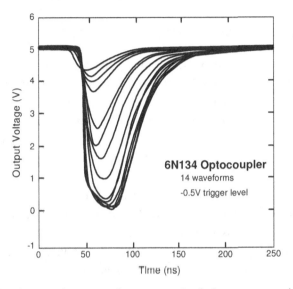

Fig. 13-6. Waveforms at the output of an optocoupler during a test run using high-energy particles from an accelerator. Waveforms from multiple particle strikes are superimposed.

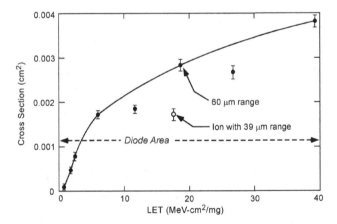

Fig. 13-7. Cross section of a high-speed optocoupler vs. LET. The cross section is significantly larger than the physical diameter of the internal photodiode because of diffusion charge that can be collected outside the boundary.

Note that the cross section is reduced for the ion with a range of 39 μm compared to that of the ion with a range of 60 μm. Less diffusion charge is produced by the particle with lower range, underestimating the cross section that is expected from long-range ions in space.

13.3.2.2 Protons

Because high-speed optocouplers can be upset with particles that have very low LET, they can also be upset from indirect reactions with protons. For conventional circuits, the proton cross section has only a weak dependence on the angle of incidence of the incoming particle because the range of the reaction products is small compared to the dimensions of the charge-collection node. That is not the case for optocouplers because of the extended area of the photodiode. Radiation tests with protons that were done at different incident angles showed a large angular dependence that also depended on energy, as shown in Fig. 13-8 [7]. The cross section is more than three orders of magnitude higher for 15-MeV protons that strike the device at large angles compared to the cross section for normal incidence, a very large increase.

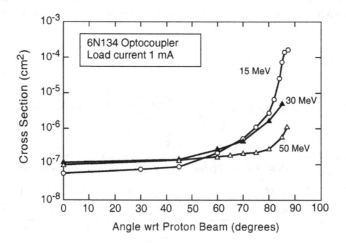

Fig. 13-8. Dependence of proton upset cross section on the angle of incidence of the proton beam [7].

Initially the angular dependence was not understood, because the tests were done without specific knowledge of the geometry of the internal photodiode. After disassembling one part, it was clear that the area of the photodiode was quite large, allowing other mechanisms to contribute to the response. Radiation measurements of charge collection using special diode structures and modeling of charge collection from protons that traveled at extreme angles through the photodetector were used to analyze the device response.

The gradual increase in cross section at extreme angles can be explained by considering the path of a proton as it traverses the photodetector. The phototransistor has a diameter of about 530 µm, 20 to 50 times larger than the size of the collection volume of an individual component in a typical silicon integrated circuit. Due to the long path length, direct ionization from protons, which can usually be ignored because of the very low LET, becomes significant. The charge from direct ionization adds to the charge produced by the recoil atom if the proton interacts with a silicon ion before it completes its path through the detector. This is shown pictorially in Fig. 13-9. The angular dependence becomes important at less extreme angles for protons with lower energy because the LET (direct ionization) is higher for protons with lower energy.

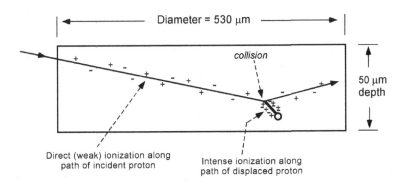

Fig. 13-9. Path of a proton through a large area photodiode showing the direct ionization path and recoil atom.

The large cross section at extreme angles causes the error rate to be about four times larger in a low-earth, high inclination orbit. It also illustrates the importance of doing proton tests at high angles of incidence, over a range of energies, as well as the importance of understanding the internal device geometry.

13.4 Optical Receivers

13.4.1 Basic issues

Optical receivers are roughly analogous to optocouplers, with much higher gain. They are also sensitive to single-event upset. Optical receivers are designed using special techniques that allow detection of very low optically induced signals, compared to optocouplers where the input signal is much higher. A simplified diagram of an optical receiver is shown in Fig. 13-10. The gated integrator allows the current pulses from the detector to be integrated during each clock cycle, subject to uncertainties in the clock cycle time. This reduces the noise, allowing the receiver to detect pulses that are near the noise floor.

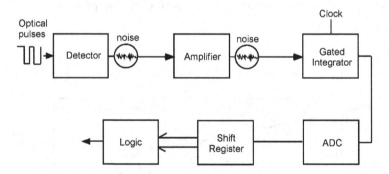

Fig. 13-10. Diagram of a basic optical receiver.

The "eye" diagram of Fig. 13-11 shows a typical signal at the output of the amplifier in an optical receiver. Detection of the correct logic signal depends on integration of this somewhat noisy signal.

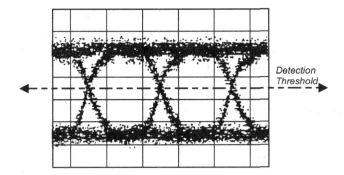

Fig. 13-11. "Eye" diagram of the amplified detector signal in an optical receiver.

The error rate depends on the signal to noise ratio:

$$\text{BER} = \frac{1}{2}\text{erfc}\left[\frac{(\text{SN})^{1/2}}{2\sqrt{2}}\right] \quad (13\text{-}3)$$

where BER is the bit-error rate, SN is the signal-to-noise ratio, and erfc is the complementary error function. Figure 13-12 shows how the bit error rate is affected by the signal-to-noise ratio. When the SN drops below 20 dB, the bit error rate falls rapidly. A common specification for bit-error rate is 10^{-9}, which requires a signal-to-noise ratio of 21.6 dB.

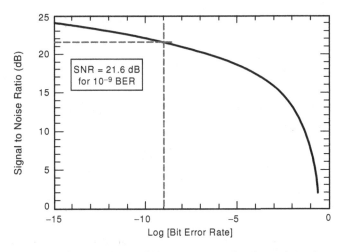

Fig. 13-12. Dependence of bit-error rate on signal-to-noise ratio.

Receiver design is more difficult at high data rates because the integration period is inversely proportional to bit rate. Various techniques can be used to increase the SNR, including coherent detection methods and using avalanche photodiodes.

13.4.2 Optical receiver radiation tests

For radiation-hardened receivers the input photodiode must be selected to have the lowest possible charge collection volume. III-V detectors, which can be designed with shallow charge collection depths because of the direct bandgap, reduce charge collection from radiation by more than one order of magnitude compared to silicon p-i-n detectors.

The results of a proton radiation test of an optical receiver are shown in Fig. 13-13 [8]. The receiver operated at 80 Mb/s. The receiver was designed to operate in a radiation environment, using an InGaAs detector with very small area. When the receiver was placed in the accelerator area it required a minimum signal amplitude of -30 dBm for acceptable operation (accelerators are noisier than typical laboratories). Radiation tests were done with various input signal levels. The BER increased slightly, even for input signals that were well above the minimum operating level. However, it was less than 10^{-9} for signals above -27.5 dBm when the receiver was exposed to the radiation beam.

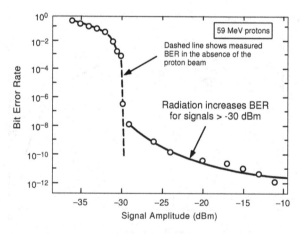

Fig. 13-13. Dependence of bit error rate on input signal amplitude during radiation tests with protons [8]. © 2001 IEEE. Reprinted with permission.

The results of Fig. 13-13 are for a radiation-hardened receiver. Commercial receivers that did not use special detectors and amplifiers would almost certainly be more affected by the presence of radiation.

13.5 Single-Event Upset Effects from Neutrons

The effects that we have discussed so far have occurred in microelectronic devices with very few internal elements. This allows us to ignore events that have very low cross sections, because the total effective area that can respond to a flux of particles is small.

The situation is very different for mainstream integrated circuits from the silicon world, where (1) the total chip area is large, often several cm^2; (2) there are very large numbers of individual transistors, particularly in microprocessors and memories; and (3) the effects of device scaling tend to make the critical charge very small.

This combination of factors has evolved to the point that many mainstream silicon integrated circuits are sensitive to upsets from terrestrial neutrons (as discussed in Chapter 8, they are created in the upper atmosphere, but a significant number reach the earth's surface). The neutrons interact in much the same way as high-energy protons, producing nuclear reactions within the device. The products of those nuclear reactions can cause upsets to occur in integrated circuits. Most semiconductor manufacturers include an estimated rate for soft errors that are produced by neutrons, as well as by alpha particles that are emitted by trace amounts of unstable atoms within the device [9].

Those same mechanisms occur in the devices that we are discussing in this chapter. They turn out to be unimportant only because the devices that we are describing are so much smaller – with so few circuit elements – compared to mainstream silicon devices.

That will no longer be the case if large-scale circuits are produced using compound semiconductor technologies. Some types of SiGe circuits may already be large enough to be affected by soft errors from neutrons and alpha particles. Generally manufacturers are aware of these effects, and design their circuits to meet overall soft error requirements.

13.6 Permanent Damage from Particle-Induced Transients

It is also possible for the ionization track from a single heavy particle to cause permanent damage. Such effects have been noted in several types of silicon semiconductors, including power MOSFETs and power transistors [10], but have only recently been reported for compound semiconductors [11,12].

In power MOSFETs, the ion track produces a temporary conduction path through the gate oxide. If the current flow is high enough, a permanent leakage path can be produced by the ion strike. That mechanism is expected to apply to compound semiconductor MOSFETs as well.

A second mechanism – single-event burnout (SEB) – is more applicable to compound semiconductors that do not use gate oxides. SEB occurs when the current in small inhomogeneous regions is high enough to cause local melting. For silicon devices, the current may be enhanced by avalanche multiplication, or by gain within the specific device structure. Although SEB can be an important limitation for power devices used in space systems, the effect usually occurs only for particles with LET >20 MeV-cm^2/mg, provided devices are not used near their maximum voltage limits (accepted derating practices require derating of at least 20% for space use).

Recent tests by Scheick, et al., have shown that breakdown effects can also take place in silicon-carbide rectifiers with high breakdown voltage [11]. Surprisingly, breakdown from ion strikes occurred at very low LET values when the diodes were biased within 5% of their maximum rated values. With 30% voltage derating, breakdown occurred at an approximate LET of 10 MeV-cm^2/mg, as shown in Fig. 13-14.

This result is potentially very significant, and could occur in other compound semiconductors used for high voltage applications, such as GaN. If ion-induced breakdown occurs near defects then the importance of the effect will depend on processing methods.

The mechanism for burnout in SiC has not been determined, but it may be associated with the high defect density that is present in SiC technology. Studies of silicon dioxide capacitors showed that defects are likely involved in ion-induced breakdown in that material, because breakdown voltage in capacitors is somewhat lower when tests are done with high fluence (this implies that the ion strikes near defects reduce the critical field for breakdown). However, unlike the results for SiC, breakdown in SiO_2 capacitors only takes place when electric fields are applied that are well above the rated breakdown conditions.

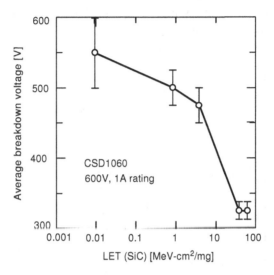

Fig. 13-14. Permanent failures of silicon carbide power diodes after exposure to heavy ions [11]. Note that these effects occur at extremely low LET values, far lower than for comparable effects in silicon structures. © 2004 IEEE. Reprinted with permission.

An additional study of catastrophic effects in SiC diodes was done by Kuboyama, et al. [12]. The devices that they studied were similar to those tested by Scheick et al., with a rated breakdown voltage of 600 V. A different experimental method was used by Kuboyama, et al., which allowed them to test devices by limiting the current, and thereby avoid full catastrophic breakdown. This made it possible to study the effects of permanent damage on a single device. They used the increase in leakage current per particle as a figure of merit to determine the threshold conditions for full catastrophic breakdown. An example of their results is shown in Fig. 13-15. The LET values for the three particles are 16.7 (Ar), 42.7 (Kr), and 73.1 (Xe) MeV-cm^2/mg.

Those results show the very abrupt increase that takes place once a critical bias condition has been reached. They used transmission electron microscopy to examine abnormal regions within the structure after catastrophic breakdown. The dimensions of localized damage sites were about 50 nm, consistent with the expected density produced by localized heating. They concluded from calculations with a device simulator program that the most likely mechanism for localized damage was trap-assisted tunneling, not impact ionization.

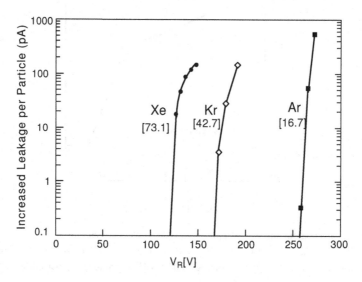

Fig. 13-15. Increase in leakage current for 600 V, 6A SiC diodes after irradiation with various types of ions [12]. © 2006 IEEE. Reprinted with permission.

The observation of catastrophic breakdown from heavy ions in silicon carbide was unexpected, particularly because of the high crystal binding energy. It is possible the high density of crystal defects plays a role in the breakdown process, which could improve as material quality advances. Even though breakdown has only been studied in power diodes, the same mechanism will likely take place in other SiC structures with high voltage ratings. The experimental results show the importance of this effect, and imply that the maximum operating voltage of SiC devices must be severely derated in order to use them in space environments.

Catastrophic breakdown has only been studied in a limited number of high-voltage SiC structures. It is possible that similar effects could take place in other compound semiconductors, including GaN. Further work is needed to improve the understanding of these effects, as well as to determine how they are affected by impurities and overall material properties.

13.7 Summary

This chapter has discussed transient effects in compound semiconductors. Upsets occur in high-speed logic devices, with threshold LETs that are very low. That is consistent with the high speed and small area of the advanced transistor structures that are used, which allows them to respond to small extraneous pulses. Fortunately, the low integration density of most compound semiconductor reduces the importance of upsets compared to silicon technology. In most cases the event rate is low, and it is possible to deal with the small number of upsets that take place with system solutions, such as error detection and correction.

Transient effects also take place in optocouplers and optical receivers. The threshold LET for upsets in those devices is also very low, and they are sensitive to upset from protons as well as heavy ions. Angular effects from protons are more important than expected in optocouplers because of the long lateral charge collection path of input photodiodes. The upset rate for optocouplers is much higher than that of individual logic devices because of the photodiode area. However, this can usually be overcome through circuit or system design, as long as designers are aware of the high upset rate.

Permanent damage from heavy ions and protons has been observed in high-voltage SiC devices, an unexpected result. Particles with very low LET can trigger this effect, and it is a very important limitation not only for the power diodes where it has been studied, but in other SiC devices as well. It is possible that other types of compound semiconductors will be sensitive to damage from heavy ions, particularly if they are used in high-voltage applications. The mechanisms for these effects are not fully understood, and may be affected by defects within the materials. It is an interesting topic for further study.

References

1. P. J. McNulty, et al., "Proton Induced Spallation Reactions", Radiation Phys. Chem., **43**(1/2), pp. 139–149 (1994).
2. B. Hughlock, A. Johnston, T. Williams and J. Harrang, "A Comparison of Charge Collection between GaAs MESFETs and III-V HFETs", IEEE Trans. Nucl. Sci., **39**(6), pp. 1642–1647 (1992).
3. D. McMorrow, et al., "Charge Collection Characteristics of Low-Power Ultrahigh Speed, Metamorphic AlSb/InSb High-Electron Mobility Transistors", IEEE Trans. Nucl. Sci., **47**(6), pp. 2262–2268 (2000).
4. G. Niu, et al., "Modeling of Single-Event Effects in Circuit-Hardened High-Speed SiGe HBT Logic", IEEE Trans. Nucl. Sci., **48**(6), pp. 1849–1854 (2001).
5. P. W. Marshall, et al., "Single-Event Effects in Circuit-Hardened SiGe HBT Logic at Gigabit per Second Data Rates", IEEE Trans. Nucl. Sci., **47**(6), pp. 2669–2674 (2000).
6. A. H. Johnston, et al., "Single-Event Upset Effects in Optocouplers", IEEE Trans. Nucl. Sci., **45**(6), pp. 2867–2875 (1998).
7. A. H. Johnston, et al., "Angular and Energy Dependence of Proton Upset in Optocouplers", IEEE Trans. Nucl. Sci., **46**(6), pp. 1335–1341 (1999).
8. F. Faccio, et al., "Single-Event Upset Tests of an 80-Mb/s Optical Receiver", IEEE Trans. Nucl. Sci., **48**(5), pp. 1700–1707 (2001).
9. R. C. Baumann, "Radiation-Induced Soft Errors in Advanced Semiconductor Technologies", IEEE Trans. on Device and Materials Reliability, **5**(3), pp. 305–316 (2005).
10. F. W. Sexton, "Destructive Single-Event Effects in Semiconductor Devices and ICs", IEEE Trans. Nucl. Sci., **50**(3), pp. 603–621 (2003).
11. L. Z. Scheick, L. Selva and H. N. Becker, "Displacement Damage-Induced Catastrophic Second Breakdown in Silicon Carbide Schottky Power Diodes", IEEE Trans. Nucl. Sci., **51**(6), pp. 3193–200 (2004).
12. S. Kuboyama, et al., "Anomalous Charge Collection in Silicon Carbide Schottky Barrier Diodes and Resulting Permanent Damage and Single-Event Burnout", IEEE Trans. Nucl. Sci., **53**(6), pp. 3343–3348 (2006).

Index

Absorption coefficient 107
Acceleration factor 127–131
Activation energy 126–128, 130–132, 135–137, 145, 147, 155, 167, 174, 183, 190, 199–201
    hot carrier 137
    MMIC 155
    table 137
Alpha particle 206, 208–210, 212, 224, 336, 351
Amphoteric doping 88–89
Annealing 79–81, 94, 96, 112, 296–298, 301–302
    contact 146
    injection 285–290, 304
    radiation damage 289, 290, 308, 317, 322–324, 327, 332, 339, 342
    thermal 250, 258–259
Arrhenius equation 127–128, 140
Auger recombination 79–81, 94, 96, 112, 296–297, 301–302
Avalanche photodiode 110–111, 306, 308, 350, 352, 341–342

Background potential 19–20
Bandgap 9–12
    band offset 27–30
    band structure 12, 13
    band tailing 76–77
    bandgap engineering 29

    direct, indirect bandgap 12, 13, 76–79
    temperature dependence 12
Beer's law 106
Bipolar transistor (BJT):
    current gain 59–61, 67
    frequency response 61–62
    gain-bandwidth 61
    polysilicon emitter 63–65
    radiation damage 247–249, 269–273
    silicon carbide 66–68
Bit error rate 349–350
Breakdown walkout 149–150
Brewster angle 87, 324
Burn-in 126, 128–130, 136, 183, 197

Carriers:
    carrier removal 249–253
    intrinsic carrier density 14, 15
    majority carriers and minority 16–18
    transport 20–22
Carrier removal 249–253
Charge density of ions in space 344, 346
Charge deposition 334, 338
Contacts:
    degradation 134, 145–156
    Schottky 184, 189
Cosmic ray; *see galactic cosmic ray*

Critical charge 338, 342–343, 351
Critical electric field 36
Critical thickness 31–32
Current collapse 54, 162–163, 165–167
Current transfer ratio 311, 314–316, 318–324

Damage factor 248, 258
Depletion width 25
Derivative plot 99–100, 299–300
Detectors:
    avalanche photodiode 110–111, 306–308
    PIN diode 108–109
    silicon diode 108, 305
    spectral width 86, 92, 106, 305
Diffusion:
    carrier transport 20
    diffusion constant 20
    diffusion length 25, 247, 314–316
Displacement damage:
    annealing 76, 245, 258–259
    bipolar transistor 247, 269, 272, 316–319
    damage constant 246, 251, 256
    damage factor 242, 248
    electron 241–242, 255–257
    energy dependence 255–258, 264
    JFET 265–266, 273
    lifetime damage 241, 244, 320
    mechanisms 240–254
    mobility 251–253
    MODFET 266–269, 276–277
    Non-ionizing energy loss (NIEL) 248–249, 254–258
    protons 248–249
    threshold energy 241–243
Diode:
    depletion width 25
    ideal diode equation 25
    ideality factor 25–26
    monitor diode 302–303

Distributions:
    Fermi-Dirac 10–11
    log-normal 119–121, 124–125, 154–155
    normal 119–122
    Weibull 122–123
Doping 16–20

Early voltage 59–61
Electromigration 134–135
Electron:
    displacement damage 241–242, 255–257
    effective mass 11–12
Electron hole pair energy 236–237
Energy band:
    band tailing 76–77
    conduction band 9–11
    transitions 13
    valence band 9–11

Failure mechanisms:
    contacts 149–150
    current collapse 162–167
    electromigration 134–135, 137
    gate lag 151–153, 166
    gate sinking 145–146
    hot carriers 148–150
    hydrogen poisoning 146–147
    sudden-beta degradation 156–157
    traps 79, 143, 145, 238, 305
Failure density function 118–122
Fermi-Dirac function 10, 11, 24
Fermi level 10–11, 16–18, 23, 33
Figure of merit for power devices 37–38
FIT rate 125
Frequency response 65–66, 68

Galactic cosmic ray:
    distribution 222–224
    origin 208–210
    solar modulation 224

Gamma ray 207–208, 217, 266–268, 276–277
Gain (optical) 85, 93–95
Gain (transistor) 65–67
   current dependence 68, 73
   radiation damage 247–248, 269–272
   sudden beta degradation 156–157
Gate:
   contact 146, 148, 158, 173, 183–185
   gate lag 151–153
   gate sinking 145–146
Geosynchronous orbit 217–218, 322
Gray (*definition*) 234
Gummel plot 61, 157

HBT; *see bipolar transistor*
Heterojunction:
   band offset 27–30
   2DEG gas 48 51
   modulation efficiency 51–52
   sheet charge density 51–52, 54
   structure 27–28
Heavy ion:
   distribution 223–224
   LET 210, 223
HFET; *see MODFET*
Hot carrier degradation 130–131, 137

Infant mortality 118, 130, 135
Intrinsic conduction 10, 15–18
Ionization:
   effects in insulators 237–238
   process 235–237
   radiation damage 237–239

JFET:
   characteristics 42–44
   cut-off frequency 44
   pinch-off voltage 43

Laser:
   catastrophic damage 173, 177, 183, 185–191, 202
   cavity 78, 91–94
   characteristic temperature 96
   distributed feedback 105
   facet damage 171–172, 179, 187–189
   five-layer 97
   frequency-stabilized 198–199
   fundamentals 91–99
   modal gain 95
   photon oxidation 187–189
   quantum-well 97–98
   radiation damage 296–302
   slope efficiency 92, 96–98, 101, 296, 299, 301–304
   spectral width 98, 102–106, 298–299
   temperature dependence 95, 98, 101, 106, 301–302
   threshold carrier density 95
   threshold current 85, 96–101, 105, 178, 185, 191, 296–304
   tunable laser 106, 196–198
   VCSEL 100–102, 194–196, 303–305
Lattice matching 31–32
Lattice strain 30–32, 36, 81, 83–85
Leakage current:
   avalanche photodiode 200, 306–307
   gate 43, 55, 273
   junction 15, 25, 29, 201–202, 352–354
   photodiode 199, 305–307
   Schottky diode 277
Lifetime:
   laser 185–190
   minority carrier 80, 89, 91, 112, 178, 246
   radiation damage 246–249

Light-emitting diode 85–91
  aging 176–183
  amphoterically doped 88–89
  dark-line defects 179–181
  heterojunction 89–91
  radiation damage 7, 283–295, 313–321
  spectral width 86, 102–103
Linear energy transfer (LET):
  definition 208
  effective LET 338
  energy distribution 210–211, 223–224, 227–228
  heavy particle 223–224, 227–228
  proton 335–336
  table 337

Material issues:
  GaAs 34
  GaN 36
  InP 35
  SiC 35
Mean time to failure (MTF) 121, 124, 135, 155, 192
MESFET 44–47
MMIC 154–155
Mobility:
  carrier density dependence 21
  displacement damage 251–253
  field dependence 21–22
  velocity overshoot 22
Modal gain 95
MODFET:
  frequency response 57
  material systems 58
  modulation efficiency 56
  operation 53–56
  pseudomorphic 56
MOSFET:
  fundamentals 55–58
  power 57–58, 160–162
  silicon carbide 57–58
  threshold voltage 55

Neutron:
  atmospheric 230–231, 351
  damage 251, 255–258, 273–274, 283
  SEU effects 351
Noise 68–70, 110–111, 348–349
Non-ionizing energy loss (NIEL):
  definition 244
  discrepancies 257–258
  electron 255–256
  neutron 258
  optocoupler 258, 321–322
  proton 255–258
Nuclear:
  reactions 230, 239, 241, 255, 335–336, 346, 351
  reactors 232

Optical receiver 348–351
  bit-error rate 349–350
  eye diagram 349
  radiation effects 350–351
Optocoupler:
  functionality 311–313
  radiation damage 313–330
  power MOSFET 327–330
  proton upset 346–348
  single-event upset 333–335, 339–351
Optoelectronic reliability:
  catastrophic optical damage 185–187
  dark defects 183
  dark line defects 185–187
  detectors 199–202, 305–307
  electrode damage 189–190
  evaluation methods 179–182, 190–193, 199–201
  facet damage 187–189
  laser diode 183–193
  light-emitting diode 176–182
  tunable lasers 196–198

Oxide:
    charge trapping 238, 273–275, 328–330
    spacer 48–51, 268, 271–272

Photodiode:
    avalanche 110–111, 200–211, 306–307
    basic photodiode 108–109, 199–200, 305–306
    PIN photodiode 108–109
Phototransistor:
    current dependence 317–318
    radiation damage 316–319
Phonon momentum 13
Photon:
    absorption 12–13, 19, 90, 93, 102–104
    extraction 87–88
    ionization 235–239, 271–272, 328–330
    photon damage 185–189
Pinch-off voltage 43–47, 56, 153
Primary knock-on atom 243
PN junctions 23–26, 86–88
Polarization charge 53–54, 277
Polytype (SiC) 35, 161
Power drift 153
Proton:
    energy distribution 215–218
    solar flare 218–222

Quaternary systems 86–89, 94
Quantum efficiency 80, 94, 105
Quantum well 48–49, 94–99, 193, 294

Rad (*definition*) 237
Radiation:
    galactic cosmic ray 206, 208–209, 235–206
    solar flare 211–212, 218–222
    terrestrial 229–232
    trapped belts 205–207, 213–217

Radiation damage:
    HBT 266–269
    laser diode 296–302, 304–305
    LED 284–295
    optocoupler 313–316, 320–330
    phototransistor 316–319
    Schottky diode 275
    transistor (bipolar) 269–273
Recombination:
    Auger 79, 96–98, 269–273, 292, 299–301
    band-to-band 10–13, 75–79
    non-radiative 12, 25–27, 86–87, 105, 176–180
    radiative 10–12, 75–79
    stimulated 78, 93–96
    spontaneous 78, 105
    surface 80, 199–201
Reflectivity 93, 104
Reliability:
    definition 118
    infant mortality 128–130
    non-constant 131–132
Reliability acceleration factors:
    current density 131, 135, 158–159, 186–187, 192–193
    electric field 130–131
    temperature 126–129, 136–137, 154–155, 197–201
Reliability mechanisms:
    basic 132–137
    contact 134–135, 145–146
    dislocations 132–134
    electromigration 134, 137
    fluorine dopant 148
    gate lag 151–153
    gate sinking 145–146
    hot carrier 137, 165
    hydrogen poisoning 162–163
    sudden beta degradation 171–173
    surface traps 147, 148, 166
Reliability screening 126–135

Sample testing 117, 138–139
Scattering 239–246
    Compton 207–208
    Coulomb 239–240
    electron 241–242
    impact parameter 239
    Rutherford 239–240
Schottky barrier 32–33, 44
Schottky contacts 189
Schottky diode radiation damage 275–278
Semiconductor:
    bandgap 13, 14
    carrier density 14–19
    direct and indirect 12
    effective mass 10–12
    fundamentals 9–23
Sheet charge 49–54, 268
Single-event burnout 352–355
Single-event upset:
    logic devices 339–343
    neutron 351
    optocoupler 343–348
    proton-induced 346–348
Single-event transients 335–339, 344, 355
Slope efficiency 92, 96–99, 101, 183, 192, 296, 299, 301–304
Solar flare 205–206, 211–212, 218–219
    distribution 206, 219–222
    modulation 224, 228
    upset rate 225–228
Solid solutions 81–82
Statistical distributions 131–137
    exponential 119
    log-normal 119–134, 136, 190–191
    normal 119–121, 124
    plotting methods 124–125
    Weibull 122–125, 186

Stimulated emission 78, 80, 83, 91–94, 183, 296
Strain and optical gain 85
Strained lattice 83–84
Surface recombination velocity 19

Temperature effects:
    laser 95–98, 301–302
    LED 312, 322–323
Ternary systems 81–83
Terrestrial radiation 212–213, 230–232
Threshold current 106–111, 329–336
Threshold gain (laser) 93
Threshold voltage 43, 45, 55–57, 145–150, 265–269
Transient:
    high-energy particles 344–346
    permanent damage 352–355
    protons 346–348
Transistor:
    bipolar 58–68, 155–160, 269–273
    JFET 42–44, 265–266, 273–275
    HEMT, HFET; *see MODFET*
    MESFET 44–47, 154–155, 162–166, 266–269
    MODFET 47–55, 145–153, 276–277
    MOSFET 55–58
    phototransistor 311–320, 324, 326
Traps:
    bulk 143, 238, 249–250, 311–320, 324, 326
    passivation layer 150, 156
    surface 19, 68, 79, 133, 143, 150–151, 158, 163–164, 166, 206, 238, 263, 271, 274, 279

VCSEL 100–102, 105–106, 194–196, 303–305
Velocity overshoot 22, 34, 36, 38

Wavelength:
　dependence on bandgap 12–13, 76–77, 81–85, 91–95
　laser diode 94, 96–100, 103–106
　LED 88, 91, 102–103

Weibull distribution 119, 122–125, 186
Wide-bandgap semiconductor 15, 17, 36–37, 160–166, 273–278